Silvia Masiero

UNFAIR

S Sage

1 Oliver's Yard
55 City Road
London EC1Y 1SP

2455 Teller Road
Thousand Oaks
California 91320

Unit No 323-333, Third Floor, F-Block
International Trade Tower
Nehru Place, New Delhi 110 019

8 Marina View Suite 43-053
Asia Square Tower 1
Singapore 018960

Library of Congress Control Number: 2024934790

British Library Cataloguing in Publication data

A catalogue record for this book is available from the British Library

Editor: Rhoda Ola-Said
Editorial assistant: Pippa Wills
Production editor: Sarah Sewell
Copyeditor: Ritika Sharma
Proofreader: Girish Sharma
Indexer: TNQ Tech Pvt. Ltd.
Marketing manager: Elena Asplen
Cover design: Victoria Bridal
Typeset by: TNQ Tech Pvt. Ltd.

ISBN 978-1-5296-2178-5
ISBN 978-1-5296-2177-8 (pbk)

To Deirdre and Lapo,
whose passion and dedication will change the world

CONTENTS

List of Tables and Figures   xi

List of Acronyms   xiii

About the Author   xvii

Acknowledgements   xix

Preface   xxiii

Statement on Names and Places   xxvii

1 **Introduction: Unfair ID**   **1**
   Digital ID: The Development Promise   2
   Unfair ID: A Data Justice Framework   5
   Organisation of the Book   8

Part 1   Identity   **13**

2 **The Digitalisation of Identity**   **15**
   What's in a Card?   16
   A Founding Principle   18
   The Architecture of Digital Identity Systems   20
   Foundational and Functional Identity: A Firm Distinction?   21
   Digital Identity and Development: Theorising the Link   22
   A Problematic Link   25
   Three Views of Digital Identity   27
      Digital Identity as a Datafier   28
      Digital Identity as a Mediator of Surveillance   30
      Digital Identity as a Platform   31
   Summary   33

3 **Digital ID: A Data Justice Framework**   **34**
   The Need for a Data Justice Lens   35
   Data Justice and Digital Identity   37
   Birth of a Framework   39
   Digital Identity: A Data Justice Framework   41
      Legal Injustice   42

Informational Injustice   43
Design-Related Injustice   45
Summary   47

Part 2   Injustice   **49**

4   Legal Injustice   **51**
A Family Matter   51
The Anatomy of Legal Injustice   52
The Numbers of Legal Injustice   54
The Politics of Anti-poverty Artefacts   57
The Biometric Artefact in Action   61
   Kenya: From Humanitarian Assistance to Double Registration   61
   Uganda: Digital ID and Exclusion From Entitlements   66
Summary   69

5   Informational Injustice   **70**
'If I Can Get Rations, It's Ok'   70
'Of What Use Is My Knowing?'   72
Opacity: The Denial of Information   74
Opacity Made Stronger: Information-Erasing Technologies   77
   Blockchain-for-Refugees: The Information-Erasing Machine   78
   Information-Erasing Databases: Ingreso Solidario in Colombia 81
On Information Orders and Injustice   85
Summary   88

6   Design-Related Injustice   **89**
What's in Design?   89
Not a Dark Side   92
Biography of a Biometric Artefact   94
Biometric Artefacts, Policing and Partnerships   98
   Data Infrastructures: Eurodac From Care to Control   98
   WFP-Palantir: Artefacts of Biometric Humanitarianism?   102
The Dark Matter of Digital ID   106
Summary   110

Part 3   Resistance   **111**

7   On ID, Solidarity and Resistance   **113**
Data Activism and ID Resistance   116
India's Right to Food Campaign: (Fair) ID as Gateway to Food   118

Reactive Focus: Opposing Aadhaar-Based Exclusions 120
Proactive Focus: ID and Advocacy for Maternity Entitlements 124
Kenya: New ID, Similar Injustice? 125
Enacting Infrastructure Justice through Deregistration 127
#WhyID: A Data Activist Lens in Practice 130
Data Activism: Towards Fair ID? 133
Summary 134

8 Imagining Fair ID **135**
Smart Cards in Tamil Nadu: Building ID Fairness? 135
The Return of Informational Injustice 137
The Lens of Infrastructure Justice 139
Imagining Infrastructure Justice: Reconstructing the
User–Provider Interface 141
Doing Infrastructure Justice: Reversing Dark Matter 144
Reversal 1: Human Rights Impact Assessments 145
Reversal 2: Anti-injustice ID Artefacts 149
Fair ID: A Conceptual Apparatus 152
Data for Dignity: Reimagining Digital Social Protection 152
Digital ID and Algorithmic Fairness 154
Fair ID in the Digital Rights Space 156
Summary 158

Conclusion: Interlocked Lenses **159**

References 163
Index 181

LIST OF TABLES
AND FIGURES

Figure 2.1: The Architecture of Digital Identity Systems   21

Figure 2.2: Digital Identity and Development: Theorising the Link   23

Figure 3.1: Understanding Unfair ID: A Data Justice Framework   41

Figure 6.1: Aadhaar-Based PDS Architecture in Karnataka   90

Figure 6.2: Point-of-Sale Machine, Karnataka, August 2014   95

Figure 6.3: PDS Godown, Karnataka, August 2014   97

Table 6.1: Entitlement to Food Grains under the Karnataka PDS (2014)   94

LIST OF ACRONYMS

AAY	Antyodaya Anna Yojana
ABBA	Aadhaar-Based Biometric Authentication
ACHPR	African Charter on Human and People's Rights
AFIS	Automated Fingerprint Identification System
AI	Artificial Intelligence
AMC	Allied Media Conference
API	Application Programming Interfaces
APL	Above-Poverty-Line
BPL	Below-Poverty-Line
CIDR	Central Identities Data Repository
CIPESA	Collaboration on International ICT Policy for East and Southern Africa
COVID-19	Coronavirus Disease 2019
DANE	National Administrative Department of Statistics
DBT	Direct Benefit Transfers
DPIA	Data Protection Impact Assessment
EIA	Environmental Impact Assessment
EMT	EUMigraTool
EU	European Union
Eurodac	European Dactyloscopy
FCI	Food Corporation of India
FIST	Financial and Accounting System
G2P	Government-to-Person
HRIA	Human Rights Impact Assessment
ICDS	Integrated Child Development Services
ICE	Immigration and Customs Enforcement
ICESCR	International Covenant on Economic, Social and Cultural Rights
ICM	Integrated Case Management System
ICTs	Information and Communication Technologies

ICT4D	Information and Communication Technologies for Development
ID4D	Identification for Development
IGMSY	Indira Gandhi Matrutva Sahyog Yojana
IIITB	International Institute of Information Technology Bangalore
IS	Information Systems
IVRP	Information Village Research Project
JAM	Jan Dhan Yojana, Aadhaar and Mobile
JSY	Janani Suraksha Yojana
LSE	London School of Economics and Political Science
MAMTA	Dr Muthulakshmi Maternity Assistance Scheme
MKSY	Mukhyanmantri Khadiyann Sahayata Yojana
MOSIP	Modular Open Source Identity Platform
NFSA	National Food Security Act
NIC	National Identity Card
NIIMS	National Integrated Identity Management System
NIN	National Identity Number
NIR	National Identity Register
NIRA	National Identification and Registration Authority
NGO	Non-Governmental Organisation
NPD	National Planning Department
NPHH	Non-Priority Households
NREGA	National Rural Employment Guarantee Act
NSR	National Skills Registry
NSS	National Sample Survey
NSSF	National Social Security Fund
PDS	Public Distribution System
PHH	Priority Households
PoS	Point-of-Sale
PRIMES	Population Registration and Identity Management Ecosystem
PRIO	Peace Research Institute Oslo
RCMS	Ration Card Management System
SCG	Senior Citizens' Grant
SDGs	Sustainable Development Goals
SDK	Software Development Kits
SIA	Social Impact Assessment

SIM	Subscriber Identity Module
Sisbén	System of Identification of Social Program Beneficiaries
TSO	Taluk Supply Office
UID	Unique Identity Project
UIDAI	Unique Identification Authority of India
UNECA	United Nations Economic Commission for Africa
UNHCR	United Nations High Commissioner for Refugees
UPI	Unique Personal Identifier
WFP	World Food Programme

ABOUT THE AUTHOR

Silvia Masiero is an Associate Professor of Information Systems at the University of Oslo, Norway. She is a long-term researcher of information and communication technology for development (ICT4D), with a focus on the role of digital platforms in socio-economic development processes. She has authored over 80 peer-reviewed research papers on topics including digital social protection, platform-mediated surveillance and decolonial approaches to information systems research. Silvia is Editor-In-Chief of the journal *Information Technology for Development*, Chair of the IFIP Working Group 9.4 on the Implications of Information and Digital Technologies for Development and a Senior Editor at the *Electronic Journal of Information Systems in Developing Countries*. She has received the Association for Information Systems (AIS) Mid-Career Award in the year 2023.

ACKNOWLEDGEMENTS

This book was born, nurtured and ultimately made possible by an amazing community of research and practice, working in and beyond the world of digital ID. With all my heart, I will try my best to do justice to such a splendid community.

The Data Justice book series at Sage has been the most helpful and supportive home the book could possibly find. Michael Ainsley has followed every step in the making of this book, and Sarah Moorhouse has been of massive support in managing all steps towards publication. As a first-time book writer, I cannot begin to tell how much their help has meant to me, and their patience taking me through the (very new for me!) process of book writing. This beautiful journey, and its current completion, would never have been possible without them.

My utmost gratitude goes to Lina Dencik, Arne Hintz, Joanna Redden and Emiliano Treré, who started the Data Justice book series and made it possible for this manuscript to become real. Big thanks all of you, and in particular, Emiliano for pitching the idea to me when all of this was just a project. While my first reaction may have been along the lines of – 'who, me, writing a book?' I have since then discovered the beauty of this form of writing, as well as the many challenges that only a great community of colleagues and practitioners has made it possible to overcome. Big thanks for believing in this project, and for giving Unfair ID a home in your series.

The book's journey began, in fact, way before it first words were written. My research interest for the injustices connected to digital ID, and for forms of resistance to them, is at least as old as my days volunteering in Palestinian refugee camps, where I first witnessed the participation of technology in direct harm on people. This book is for all the refugees, humanitarians and activists whom I met while volunteering in refugee camps in the West Bank, Lebanon and Jordan in the years 2008–2013. As Palestine is subject to one of the deadliest military assaults in history while I write these lines, it is all its people that I have in mind.

Writing my PhD thesis 'Imagining the state through digital technologies: A case of state-level computerization in the Indian Public Distribution System' marked a substantial part of the journey. My PhD fieldwork brought me to the Indian state of Kerala, and to its people's work and activism for food security. It was in Kerala that I first encountered the participation of digital ID in the Public Distribution System (PDS) on which much of my research has focused, and in the neighbouring state of Karnataka that I could follow the evolution of the same system towards biometric verification. Sunil Mani hosted me at the Centre for Development Studies Trivandrum during my PhD research, and work with Amit Prakash from the International Institute of Information Technology, Bangalore, has enabled the evolution of my research. Special thanks go to Soumyo Das, with whom I developed the data justice framework that evolved into this book. Friendships born from research on the PDS are strong and splendid to this day: Bidisha Chaudhuri, Janaki Srinivasan, Balaji Parthasarathy, Tarangini Sriraman, Grace Carswell, Geert De Neve, Savita Bailur, Rajesh Veeraraghavan, this work would never have been imaginable without your inputs. More friendships built while conducting the same research – I am looking at you Aleksi Aaltonen, Attila Marton, Elena Parmiggiani, Florian Allwein, Kari Koskinen, Marta Stelmaszak, Margunn Aanestad, Sajda Qureshi and Katherine Wyers – have deeply nurtured the researcher and person I have become. I will never be grateful enough for that.

A massive special mention is for Shirin Madon, my PhD supervisor from back in the days and an irreplaceable friend and colleague to this time. I cannot count the times Shirin encouraged me to persist in my research when I was close to dropping everything, to question my logic, and to keep up with the work as my academic journey continued. Having moved to Norway from the United Kingdom only moved online our beautiful coffee conversations, which stand as the most wonderful occasions I have for receiving wisdom on research and life. Thank you Shirin, you absolute legend.

And then this book. I like saying that it was written from 'being in the world' of digital ID; it is hard to count the workshops, symposia, conferences, coffee conversations, articles read, debates encountered, communities met, and especially, people who made part of it. To start with, Malavika Raghavan and Keren Weitzberg have made it possible to gather together an amazing group of people with an interest in digital ID research at the London School of Economics and Political Science (LSE) in March 2023, where many ideas contained in this book's pages have been discussed. Malavika's input on the legal aspects of biometric identification has been invaluable, and so has been Keren's on the history of refugee registration in Kenya, illuminating me on the intricacies of double registration. Malavika, Keren – thank you! It is friendships like ours that nurture every bit of my life and work. Interactions with Frank Hersey, Eve Hayes de Kalaf, Edgar Whitley, Fabio Cristiano, Emrys Schoemaker, Carolina Polito, Gianluca Iazzolino and Bruno Oliveira Martins have deeply enriched the argument that the book makes.

This book has involved venturing into research on ID artefacts that I had not encountered before. Margie Cheesman has introduced me to the use of blockchain in digital humanitarianism, thanks to which the notion of *information-erasing artefacts* contained in this book has been theorised. Hers is also the notion of *infrastructure justice* used here to imagine new forms of fair ID. Joan López has informed me on Colombia's Sisbén, and the inner workings of algorithmic welfare that lie behind its outcomes. As I was approaching the end of this manuscript, meeting Divij Joshi has illuminated me on the functioning of another social welfare algorithmic system, Samagra Vedika, pointing me to the intertwining of the discourse of digital ID with that on algorithmic justice. To all of you, big thanks.

Members of the Global Data Justice team at Tilburg University have been of massive inspiration to the theoretical framing of the book, starting with Linnet Taylor whose definition of data justice is the book's guiding light. Aaron Martin's input on the book proposal, as well as several draft chapters, has been of massive support, as has been our friendship of many years. Gargi Sharma, Siddharth de Souza and again Joan López: big thanks for your inputs, and for great conversations as we finally managed to meet in person! Same goes for Katelyn Cioffi, whose inputs on the human rights perspective on digital ID, and on the case of *Ndaga Muntu* in Uganda, have been of massive help. And for Azadeh Akbari, with whom I share the journey of the edited volume Critical ICT4D. Your friendship, your surveillance studies work, and your critical perspective inspire me every day.

This book would not exist without work on advocacy for people experiencing ID-induced harm, and without the civil society organisations thanks to which it is even possible to imagine fair ID. Mariam Jamal has taken me through the important work of Haki na Sheria in Kenya, illuminating me on the practicalities of legal and humanitarian advocacy. Big thanks to Tom Fisher, Giulio Coppi, Antonella Napolitano and Marianne Diaz for many, illuminating insights on the landscape of civil society work on digital ID. Stefania Milan and, again, Emiliano Treré have introduced me to the notions of data activism through which the book proposes to imagine fair ID: your friendship, and our fantastic collaboration to the edited volume 'COVID-19 from the Margins: Pandemic Invisibilities, Policies and Resistance in the Datafied Society', inspire me massively. First came your invitation to be part of the COVID-19 from the Margins project back in 2020, and then the many adventures we are off to together! I am so lucky to have you, amazing colleagues and friends, in my life. All errors in this text are my own.

As these thankful notes may illustrate, research creates connections and friendships that may last a lifetime. Not for a minute has my journey been lonesome. My team at the University of Oslo has been the most supportive environment to complete this work, and has offered a great space to pitch its ideas and test their theoretical robustness. My head of group Petter Nielsen has given me the most amazing support in finishing this book, and I cannot thank him enough for that! Tejas Kotha, my co-host in the Sociotechs podcast, has been a splendid proofreader and the loveliest moral support I could imagine to have. All my friends from research and life, with a special mention to my book-writing peer Marco Marabelli, have given me the positive energy to make it through this journey.

Writing my first book taught me a lesson. A book is not written alone; it is the product of many conversations, confrontations, even vivid debates with the community it contributes to. It also is a means to make academic work accessible across activism and practice, and in the case of this book, to persons of my life that do not belong to my research sphere. This book is for my parents Anna and Roberto, who taught me the passion for human rights that inspired the whole project. Every word of Unfair ID, my first book, is for them.

Published in collaboration with the Data Justice Lab as part of the Data Justice Series.

PREFACE

In December 2010, I embarked onto my first PhD field trip to the Indian state of Kerala. The purpose of my PhD was to study the computerisation of India's Public Distribution System (PDS), the largest food security scheme in the nation. The PDS is an essential source of livelihood for millions of people, providing food and non-food items through ration shops around the whole country. Computerisation, back then proposed by the Kerala government in the form of automation of several phases in the PDS supply chain, was put forward as a route to streamlining social protection, in a nation affected by systematic leakage of PDS items to the market and away from the poor.

Animated by a willingness to understand PDS computerisation, I was surprised to encounter the reaction from one of the first government officers I interviewed. 'There's very little tech in that programme', he said. That was true from his end: the electronic system, or e-PDS, through which computerisation was happening was largely invisible to both users and non-users, with technology applied in back-end stages of allocation and distribution of goods to ration shops. In the broader citizenry, people showed surprise that I was even studying that system, in which, they said, there was little or no technology to research. A personal acquaintance candidly asked, 'why is LSE (the institution to which I was affiliated) even funding this study?'

What was happening at the same time is now history. In September 2010, the Unique Identification Authority of India released the first Aadhaar number, a number deemed to become profoundly entrenched in the lives of India's residents. The 12-digit unique number, obtained upon a process that captures the essential biometric and demographic data of enrolees, is presently ubiquitous in India's governance machine. As noted throughout the literature, an Aadhaar number is a lot more than the number itself; it is the epitome of how a digital identity, converting the individual into machine-readable data, has become central to national governance. With its enactment of a 'biometric state' (Breckenridge, 2014), Aadhaar is transforming not only the very access to PDS ration shops, but that to the large plethora of social protection schemes in the nation.

In its essence, Aadhaar has become a way to do development. Grown to be the largest programme of its kind in the world, Aadhaar is the blueprint on which more countries, building on a novel orthodoxy of digital identification for development, plan to datafy their population and its registries. By doing so the Aadhaar model has come to redesign not social protection alone, but the very ensemble of state–citizen relations predicated, argue theorists of the biometric state, on the 'legibility' of populations. It is that legibility, the identification for development orthodoxy goes, that enables service provision for large masses of poor and undocumented people, otherwise invisible to the state and humanitarian agencies.

But more stories come to mind of my field days. It is 2018, and I sit in front of a ration shop in the state of Karnataka, which is undergoing a complete transition of PDS to Aadhaar-based recognition of users. Here, in the queue outside the shop for collection of monthly rations, my co-fieldworker and I witness entire households standing in line, in the hope that at least one member of the family will be able to authenticate through their fingerprints. Those who cannot authenticate after three failed attempts are invited to come back at another time. We hear stories of rations denied for months in a row, the media report of hunger deaths due to people being excluded from the biometric food security system. To say that with Corbridge et al. (2005), the light of 'development' is very dim in these stories, and echoes the narratives on injustice that have started to emerge on datafied anti-poverty systems, increasingly in relation to digital identification.

This book is a journey into these injustices and their import. Its argument is that there are forms of injustice – of legal, informational and design-related types – that are produced *specifically* with digital identity systems, and that could not be produced without their advent. In studying unfair outcomes of digital identity, Taylor's (2017: 1) notion of *data justice* as 'fairness in the way people are visualised, represented and treated through their production of digital data' is the book's guiding light. In a world of technologies of governance built to convert populations into machine-readable records that make them amenable to administrative purposes, the data justice lens shines light on the *people* at the centre of these datafying efforts. With digital identity systems spreading across countries, spurred on by global initiatives linking ID to development, the book offers a journey throughout unfair ID, exploring routes to the production of injustice and to the construction of people's resistance against it.

In the increasingly polarised debate around digital identity systems, this book comes with two caveats. The first is that the book comes at a time of increased critique to digital identity systems, especially when related to the 'development' outcomes that they are claimed to produce. In the upsurge of studies of exclusion and human right violations connected to digital identity, a data justice lens offers a guiding light to recognise the anatomy of the phenomenon. It is with this light, pointed to the experience of unfairness lived by digital identity users, that the book articulates its journey, answering the question, *what happens to the user* as a result of digital identity?

A second caveat is that a book on unfairness of digital identity could – by its definition – have reversed the lens and been a book on fairness. It partially is; the book's journey reflects on how, taking stock of injustice, new routes to fairness can be imagined in digital ID. The book does, however, align with the Centre for Human Rights and Global Justice (2022) of New York University in recognising the same mission of digital ID research; that of standing with the vulnerable and marginalised, and researching the mechanisms and practices through which they are rendered so. The chapters of this book reach to the core of the production of marginalisation, and by doing so, afford imagining forms of digital identity in which the power goes back to the datafied subject. While being a book on unfair ID, it is a book with a look to the future, a future of potential new justice for the digitally identified.

As a result, this book is a hymn to hope. It is the hope that, by becoming more aware of which injustices are produced by digital identity and how, more just forms of identity can be built. Escaping datafication of the individual is a hardly imagined effort; transforming it into fairer forms is a lot more thinkable, and a programmatic intent of this research. It is in pursuit of fairness, and of its production in a world increasingly mediated by digital identity, that this book is written.

STATEMENT ON NAMES AND PLACES

Names of individuals encountered during my field research are pseudonyms, except when referring to public servants or otherwise public personalities speaking about their official role. Names of places and organisations, where used, are real.

1

INTRODUCTION: UNFAIR ID

In this book you will encounter many stories. They are traversed by technologies, places, social policies and information systems, but they have one main thing in common. All are stories of people.

You will encounter the story of Aisha, who since many months is waiting in vain for a *ration card,* the document that her family needs in order to access vital food grains. That of Ayanka, whose fingerprint is not recognised by the biometric machine at the local ration shop, and who therefore risks being denied basic food rations to feed her household. The stories of Kenyan nationals registered as refugees in their own country, and those of displaced persons whose data, upon registration as asylum seekers, have become searchable in EU police force databases. The stories of people denied basic food subsidies as a result of an algorithm erroneously flagging them as car owners. You will meet many people, whose stories, across time and space, have a common denominator: all are traversed by digital identity systems, which have deeply affected core aspects of their lives.

Digital identity systems convert individuals into machine-readable data. They are defined by the digital performance of three functions (Nyst et al., 2016: 8–9): they *identify* individuals, meaning they establish information about them; they *authenticate* individuals, which involves asserting an identity previously established during identification; and upon successful authentication, they *authorise* individuals to access services or goods that may be of crucial importance to their lives. The below-poverty-line person who uses their biometrics to access food rations has their credentials verified by the system, which has stored their data during identification, and only upon successful authentication can they be authorised to access the food grains, sugar and pulses that their state-subsidised ration consists of.

Digital identity systems are extremely prominent in modern societies. According to the Identification for Development (ID4D) Integration Approach study, of 198 surveyed countries only 8% had no digital ID into place (World Bank Group, 2015: 10). Identification has become so intrinsic to the practices of the modern age that any attempt to explain its importance feels, in a way, reductive; the fact is *many* core processes in human lives are mediated by identification. Consider, for example the person who needs to prove their poverty status to access social cash transfers, the refugee who registers with a humanitarian organisation to receive food, shelter and essential provisions or the homeless person that can escape police violence only by producing proof of identity. Identification, with its mediation through the digital systems studied in this book, is central to the most vital practices of their lives.

The embodiment of identification with narratives of securitisation and surveillance has deep roots in social and information systems research (cf. Lyon, 2001, 2009; Whitley & Hosein, 2010; Bennet & Lyon, 2013). And yet, notes University of Virginia researcher Aaron Martin (2021: 1), over the last decades a substantial turn has taken place: a *development turn*, associating identification to the achievement of human development goals has become central to the narrative around digital ID. Motivating large-scale efforts to bring together the demand and supply of high-end identification technologies, the development turn of digital identity has given a human face to technologies originally conceived for policing and profiling. The promise underpinning this turn, which embeds an orthodoxy of 'digital identity for development', is the starting point of this book, and one that will inform the conversation it engages with people whose rights and entitlements have come into question after ID digitisation. How did 'identity', with its digital embodiments, and 'development' become part of the same narrative?

Digital ID: The Development Promise

Low and middle-income countries are seeing wide diffusion of digital identity schemes. Aadhaar, the world's largest digital ID scheme enrolling over 1.2 billion residents of India (UIDAI, 2022) was launched as the Unique Identification Project (UID) in 2009. With the Modular Open Source Identity Platform (MOSIP), established in 2018 at the International Institute for Information Technology Bangalore (IIITB), India, over 100 million digital and civil identities have been issued across all partner countries (McDonald, 2023). Humanitarian agencies such as the United Nations High Commissioner for Refugees (UNHCR) are also adopting digital identity systems, making them central to the provision of assistance to refugees and displaced persons (Madon & Schoemaker, 2021; Sandvik, 2023). The spread of digital ID in contexts of systematic marginalisation is a trend in expansion, and a characteristic of the present times.

What are the reasons for such a trend? To understand that, we need to unpack the promise made by the recent, powerful orthodoxy of identification for development. This is an orthodoxy that inextricably links identification to human development goals, creating the epistemological basis for the diffusion of digital ID systems in vulnerable contexts.

That of identification for development is an openly articulated orthodoxy. Even if its foundations were already spelled out, for example in early speeches by India's Aadhaar founder Nandan Nilekani (2013), the vision has been openly codified with multiple initiatives around the globe. Funding released by ID4D, the World Bank initiative currently supporting ID and civil registration systems in 58 countries and four regional organisations, ties in with achieving Sustainable Development Goal (SDG) target 16:9 to 'provide legal identity for all including free birth registrations by 2030'. But identification, notes former ID4D coordinator Mariana Dahan in work published with Alan Gelb from the Centre of Global Development, is also a crucial enabler of a plethora of targets across the other SDGs (Dahan & Gelb, 2015).

The theoretical link between legal identity and development is extensively discussed in digital identity literature. Central to such a link, and a conceptual basis for target 16:9, is the connection between access to services and the legibility of individuals, from the state or at large from service providers. In their 2015 working paper, Dahan and Gelb unpack such a link: if the unidentified person, invisible to the service provider, is to obtain legal identity, the provider can recognise them as an entitled subject, which is the basis to provide the services they need. It is this legibility, established in terms of legal identification, that makes the individual a subject of entitlements (Dahan & Gelb, 2015).

This idea has grown in prominence over time. The first version of this chapter was written shortly after the 2022 Annual General Meeting of ID4Africa, described on its website as 'an NGO Movement that accompanies African nations on their journeys to develop robust and responsible identity ecosystems in the service of development and humanitarian action'. The Executive Chairman of ID4Africa, Joseph Atick, opened the meeting with a central statement: the importance of ID has shifted, from being based on identity alone to identification-enabled service provision. Atick described the shift, and ID4Africa's work in relation to it, as follows:

> (...) our end goal in ID [for development] is not about digital identity, it is about building public infrastructure for governance and service delivery as frictionless, robust and respectful of people's rights and liberties – including the right to have legal identity. This is our updated objective. This is definitely more sophisticated than traditional identity management, and unless it is done right and quickly, countries will fall behind. (Quoted in Hersey, 2022: n.p.)

This point illuminates another component of the digital ID for development orthodoxy, namely the centrality of biometric technologies in development-oriented identification. The purpose of digital identity systems is that of establishing the *uniqueness* of a person's identity, which biometric technology, with its capacity to associate a person's identity to their biological features, is designed to do. Technology vendors occupy a prominent position in digital ID narratives, and one that, research notes, especially needs illumination (Martin & Taylor, 2021b; Polito & Alaimo, 2023).

As recently noted in civil society work (Access Now, 2024), narrations of digital ID need greater focus on the providers that make identification technologies available and commercialisable across markets.

It is on this orthodoxy, with its belief in connection between identification and human development, that the development turn of digital identity is constructed. At the core of digital identity is the ability to match subjects with their entitlements: through digitisation of their demographic and, increasingly, biometric records, people can receive the exact entitlement they are promised under governmental, private or humanitarian schemes. In this respect, digital identity systems are twofold: they convert individuals into digital data, matching such data with people's existing entitlements. The architectural design of digital ID technologies carries in itself the basis of its development promise.

For years, this orthodoxy has underscored the link between digital identity and a twofold development purpose. Work on transformative social protection by Devereux and Sabates-Wheeler (2004) helps unpacking this logic: on the one hand, including all those entitled to a given service affords fighting an *exclusion error*, which takes place where entitled users are excluded from service provision. On the other hand, it affords excluding all the non-entitled to that service, fighting an *inclusion error* which causes risk of resource depletion, diversion and leakage away from targeted populations. Being able to include all and only those entitled to humanitarian provisions, state services and government schemes points to the essence of the promise of digital identity, which animates the adoption of ID technologies across nations, organisations and humanitarian agencies.

And at the same time, the orthodoxy has started to crumble. The stories collected in this book point to a central theme: through the years, while inscribed in development policies, digital identity systems have produced detrimental outcomes that have strongly impacted the lives of users. Over time and across geographies, severe issues have emerged with them: made conditional to digital authentication, access to basics such as food, healthcare and social protection have been denied to non-authenticated users, even if genuinely entitled (Khera 2017; Muralidharan et al., 2020; Centre for Human Rights and Social Justice, 2021). The interoperability of digitally produced databases has made the records of asylum seekers searchable across police authorities (Pelizza, 2020; Martins et al., 2022). The opacity of governments on how user data, stored in digital identification databases, have been used to concede or deny essential services has generated mistrust and concern in recipients (Whitley et al., 2014; Chaudhuri, 2021; Masiero & Bailur, 2021).

This book offers a journey into these ruptures. With its data justice lens, it studies injustices in digital identity systems *as experienced in the lived reality* of individuals affected by them. In this sense, this book differs from a large body of digital ID work: in a world of reports about funders, providers and implementers, this is a book about *people* that live the journey of digital identification. It is due to this book's focus on people, as the receiving end of digital ID efforts, that I chose a lens of *data justice*, a perspective centred on human beings and on the consequences of their production of digital data.

Unfair ID: A Data Justice Framework

The argument of this book is that there are forms of injustice that are produced *specifically* with digital identity systems, and that could not be produced without their advent. The reason for this is that these injustices require a specific architecture to be created: it is, in fact, the architecture of digital identity systems that is intimately linked to their production. To understand such an architecture, we need to delve deeper into some key terms of digital identity research, which we will encounter frequently through this book.

As noted above, and as recognised in Nyst et al. (2016: 8–9), digital identity systems involve the digital performance of three interlinked, but different functions, defined as follows:

- *Identification*, which refers to the process of establishing information about an individual. This involves 'examining "breeder documents" such as passports and birth certificates, consulting alternative sources of data to corroborate the identity being claimed and potentially collecting biometric data from the individual'.
- *Authentication*, which refers to the process of asserting an identity previously established during identification. This involves 'presenting or using an authentication credential (that was bound to the identity during the identification process) to demonstrate that the individual owns and is in control of the digital identity being asserted'.
- *Authorisation*, meaning 'the process of determining what actions may be performed or services accessed on the basis of the asserted and authenticated identity'. This function determines, or denies, the possibility of accessing goods or services as a result of successful authentication.

A central feature of the architecture of digital identity systems is that the functions of identification, authentication and authorisation are linked together (Nyst et al., 2016). Specifically, digital identity systems are based on the conditionality of authorisation to successful authentication, which in turn requires that an individual has been previously identified. The below-poverty-line person that goes to collect their food ration, in virtue of the fact that the system (a) knows their identity and (b) associates that identity to a status of poverty, goes to a ration shop where they authenticate with their fingerprint through a biometric point-of-sale machine.

If their fingerprint is recognised, based on previous enrolment of the person in the system, this leads to the authorisation of ration disbursal; if authentication is instead unsuccessful, authorisation to disbursal is to be denied.

A central point here is that a data justice lens is a potent epistemic tool to study this architecture. This book is guided by the definition of data justice developed by Linnet Taylor, professor of International Data Governance at the University of Tilburg. Taylor (2017: 1) defines data justice as 'fairness in the way people are visualised, represented and treated through their production of digital data'. The suitability of a data justice lens to study digital identity systems stems directly from how these systems produce digital data, through which people can be treated in more or less fair ways. Just like an idea of justice, Taylor (2017: 1) continues, is needed for the establishment of the rule of law, an idea of data justice, centred on people's production of digital data, is needed to 'determine ethical paths' in a datafied world, paths that directly intersect the lives of digitally identified people.

While the Oxford dictionary defines *fairness* as 'impartial and just treatment and behaviour, without favouritism and discrimination', both the concept itself and its relation with justice have been largely debated over time. In his essay 'Justice as Fairness', John Rawls (1971) develops a conception of justice articulated on two principles: 'first, each person participating in a practice, or affected by it, has an equal right to the most extensive liberty compatible with a like liberty for all; and second, inequalities are arbitrary unless it is reasonable to expect that they will work out for everyone's advantage, and provided the positions and offices to which they attach, or from which they may be gained, are open to all'. He states 'elimination of arbitrary distinctions' as fundamental to justice: it is arguably from the second principle on the arbitrariness of inequalities that the notion of *distributive justice* emanates (Dencik et al., 2022). At the same time, the principles are not meant to be separate from each other, but to point to a unitary conception of justice as fairness.

On the other hand, the feminist political philosopher Nancy Fraser (2008) authored one of the most fundamental problematisations of the very idea of distributive justice. In her essay 'abnormal justice', Fraser moves away from focus on the distribution of goods in a society: her attention shifts on to the 'grammar' of justice, meaning the very conditions according to which justice is understood and made sense of. The point, Fraser continues, relates to three different *nodes* of abnormality: the 'what', the 'who', and the 'how' of justice, to be regarded as 'persistent features of justice discourse for the foreseeable future' (Fraser, 2008: 419). As noted in Dencik et al. (2022: 128), Fraser's framework helps consider how datafication intersects the three nodes of abnormality, leading to interrogating not only what is fair, but also, and especially, *who* establishes criteria of fairness and *how* these are enacted in practical terms.

Flowing from Taylor's (2017) definition, a 'fair ID' system may produce a non-discriminatory way to visualise, represent and treat the identified persons. A 'fair ID', read in data justice terms, should grant justice in key aspects of people's datafied existence, pertaining to the ontological questions that Fraser (2008) highlights. Seen in this light, fairness comes across as a necessary condition for accomplishing the human development goals reflected in the orthodoxy of ID for development. Where distributive justice embraces the outcomes of societal allocations of resources, a data justice framework impinging on Fraser (2008) leads to question 'the very conditions that underpin how justice is understood, debated and advanced' (Dencik et al., 2022: 128).

And yet, large research evidence exists on the detrimental outcomes of digital ID. The below-poverty-line individual who cannot authenticate biometrically and sees their food ration denied; the person rendered stateless by a hostile legislation reflected in identification technologies; the asylum seeker whose data are made accessible to police authorities across borders, all provide detailed narration of the detrimental impact that digital ID systems have on people's lives. In the mass of empirics on unfair ID, grouping together injustices of different types, we need a guiding light for a systematic analysis of the lived experience of users. It is here that data justice provides the light we need.

This book argues that studying digital identity systems from a data justice angle leads to revelation of three types of injustice. These types are different, but grouped by an important feature: all of them are fundamentally predicated on the architecture of digital identity systems. First theorised with the data justice framework put forward by Masiero and Das (2019), the injustices that this book examines are described as follows:

> *Legal injustice.* This type of injustice stems from the conditionality of legal rights and entitlements to digital identification and authentication of individuals. Legal injustice leads to the loss of universality of fundamental rights: this is for example the case for the person whose food ration is denied by failed authentication, who loses their right to food as a result of that. And it is the case for the person that, deprived of their citizenship by law codified into technology, is rendered stateless and denied legal rights. Built on failure of the authentication-authorisation nexus embedded in digital identity systems, legal injustice deprives the individual of rights and entitlements fundamental to them.

> *Informational injustice.* This concept refers to the obscuration of essential information on how data provided under digital identity schemes are handled and used by service providers. This is what happens, for instance, when the assignation of government subsidies to individuals or households is based on opaque cross-checking of personal data of residents, which the

state already holds. This is also the case of users of digital identity systems who are not informed, or have been misinformed, on the real purposes of collection of their data by providers, for example when such data are collected to enable the transition from in-kind to cash transfer social protection programmes (Masiero & Arvidsson, 2021). Be it caused by misinformation, or by lack of any information, cases of informational injustice abound across the digital identity systems studied in this book.

Design-related injustice. This is a type of injustice that results from digital identity systems whose architecture is designed in ways that cause harm to the user. The concept draws on Costanza-Chock's (2020) theorisation of how injustice can be inbuilt in technology design and systematically perpetuated through the technology itself. Gender binary identification systems that, identifying the individual as necessarily female or male, impose binarism as a condition for existence are central to the concept. So are technologies that racialise the individual, creating false positives and negatives due to the assumption of a default body (Rao, 2013; Browne, 2015). Design injustices also emerge with the formation of gaps between how a digital identity system is designed and the needs of people, including protection of one's identity in contexts where disclosure puts the person in peril.

Diverse as they may be, all three types of data injustice are produced within the space of digital identity systems. And crucially, all three injustices pertain to the individual's production of digital data, hence qualifying them as data injustices to be studied in relation to digital ID architectures.

A caveat is that this taxonomy is surely not the only one possible, and does not seek to be exhaustive in terms of the types of injustice that digital identity systems can produce. But the fact that it impinges on digital identity system architecture, studying injustices that are intrinsic to digital ID, makes it the intellectual device that will guide this work. Large amounts of empirics make it hard to order and classify the different forms of harm associated to digital identity systems; based on the central features of these systems, this taxonomy offers a way to do so.

Organisation of the Book

This book is articulated along three sections. I begin the first section, *Identity*, with a brief synoptic perspective on the notion of identity and its digitalisation. In Chapter 2, I first discuss the essential precondition – identification as a requisite for receiving entitlements – that led to the first conception of digital identity systems. I then offer a taxonomy of views of digital identity systems, noting how the literature has alternatively classified them as datafiers, platforms or mediators of surveillance, and examining the implications of these differences in perspective. Relying on cases from across countries, the chapter prepares the ground for discussing digital identity systems throughout the book.

Chapter 3 introduces the framework through which the book is structured. The book's object is studied through a data justice lens, applying Taylor's definition of data justice to digital identity systems. Having offered a rationale for doing so, the chapter introduces the dimensions of legal, informational and design-related injustice along which the framework is articulated. The chapter also puts the framework in the context of existing data justice research and highlights its relation with other taxonomies that offer alternative ways to identify, map and distinguish different dimensions of data justice.

In the second section, *Injustice*, I use my data justice framework to articulate the book's journey across the injustices of digital identity systems. In Chapter 4, I focus on legal injustice, which stems from the conditionality of legal rights and entitlements to digital identification and authentication. Establishing the conditionality of these to the authentication-authorisation nexus results in the loss of universality of rights as fundamental as the right to food, which the chapter illustrates through my research on the incorporation of India's Aadhaar in the Public Distribution System (PDS), the nation's largest food security scheme. The argument is extended with international cases including Kenya, where the issue of double registration denied the right to an ID card to many nationals applying for it (Weitzberg, 2020a, 2020b; Haki na Sheria, 2021), and Uganda, where the conditionality of healthcare and cash entitlements to digital ID has resulted into large-scale exclusions of entitled people (Centre for Human Rights and Global Justice, 2021). It is through the combination of my own research data with the work of fellow digital identity researchers that my illustration of legal injustice is made.

Chapter 5 focuses on informational injustice, which is perpetrated by providers of entitlements by obscuring information on how data within digital identity systems are used. Informational injustice stems, in the first place, from people's inability to interrogate use of their data within digital identity systems: this was the case for my interviewees in southern India, faced with the alternative of enrolling with the national biometric architecture or becoming unable to access essential social protection schemes. The chapter combines my field stories, narrating PDS users' preoccupation with a possible transition from food subsidy to cash transfers, with information-erasing artefacts placed, respectively, in a blockchain-based humanitarian scheme in Jordan (Cheesman, 2022a, 2022b, 2022c) and a subsidy system built for vulnerable households in Colombia (López, 2021a, 2021b). I again use my fieldwork on India's PDS to introduce the core concept, then corroborating it with examples from digital identity systems across nations.

Chapter 6 completes the section by focusing on design-related injustice. Drawing on the work of MIT researcher Sasha Costanza-Chock (2020), whose concept of design injustice shows how injustice can be built in technology design, the chapter reviews the stories of multiple groups of digitally identified subjects harmed, in different ways, by system design. In my fieldwork in India, design injustice showed in the form of a system that, while focusing on excluding all the non-entitled to food rations, offered no provisions to include those that, while being genuinely entitled, could not authenticate. Putting the concept in international perspective, the chapter covers stories of asylum seekers put into peril by the sharing of their data with police authorities (Pelizza, 2020; EuroMed Rights, 2023), the partnership of the World Food Programme (WFP) with Palantir, whose software has been associated to large-scale human right violations (Amnesty International, 2020), and concludes using Iazzolino's (2021) concept of 'infrastructures of compassionate repression' to conceptualise systems that conflate care and policing in the same digital architecture. The heterogeneity of examples used in the chapter paints a comprehensive picture of the forms that design-related data injustice can take.

The journey from injustice to resistance that animates this book is the reason why the third section, *Resistance*, is positioned right after the examination of injustices. In Chapter 7, I focus on how resistance to unfair ID is conceived, planned and enacted by the people. The chapter relies on Milan and Van der Velden's (2016) notion of data activism as 'the range of sociotechnical practices that interrogate the fundamental paradigm shift brought about by datafication': stories from the Right to Food campaign in India, the work of human rights' organisations towards restoration of the rights of double registered people in Kenya and the international #WhyID campaign coordinated by Access Now enact a data activist lens in narrating many practices of resistance to unfair ID. It is through these practices and their constructive potential that new forms of 'fair ID' can be imagined, based on respect for and indeed enhancement of human rights.

The exercise of imagining forms of fair ID is performed in Chapter 8. The chapter begins with a view of the journey from injustice to resistance that informed the book, illuminating how the interplay of data justice and data activism allows imagining new routes to ID fairness. It then illuminates the concept of infrastructure justice (Cheesman, 2022a) as essential to imagining such routes, theorising not only how injustice takes place, but the core dimensions along which it can be overcome. Redesigning artefacts and enabling mitigation frameworks are presented as routes to build fairness in digital ID: illustrated with empirical material, both routes are seen as means to take stock of injustice, building on it for fair ID systems to become realisable.

As noted in the opening, this book is a hymn to hope. At its heart is the tenet that only through a critical journey across forms of ID unfairness, which examines their core manifestations and the foundations behind them, can fairness in digital identity be imagined and subsequently planned. This book's exploration of injustices, appearing in multiple aspects of digital identity, comes with examples of how people have opposed injustice, thereby imagining forms of identity built and made to promise fairness to their users. It is with the willingness to contribute to a new world of fairness, based on consciousness of a current state of unfairness, that we embark on our journey through unfair ID.

PART 1
IDENTITY

2

THE DIGITALISATION OF IDENTITY

In 'Seeing the State: Governance and Governmentality in Contemporary India', Stuart Corbridge and colleagues (2005) engage the argument of 'Seeing Like a State: How Certain Schemes to Improve the Human Condition Have Failed', James Scott's (1998) study of governmental schemes in terms of the legibility of populations. Turning Scott's perspective upside down, Corbridge et al. interrogate how *populations* see the state, enquiring how and where 'sightings of the state' are produced in people. Such sightings emerge from *encounters*, meaning the moments in which people 'meet' the state in its physical manifestations. As such, the state is no abstraction; it is the policeman patrolling the streets, the officer delivering ration cards to recipients, the ration dealer who disburses or denies food rations to people. Crucially for this book, sightings of the state are deeply impregnated with technologies, which – like the fingerprint reader on which authorisation is based or the eye scanner through which refugees authenticate to receive entitlements – structure the encounters through which people see it.

The lens of 'seeing the state' through physical encounters inspires the whole book. Its starting point is a simple one: it is at the physical interface, where the person encounters the service provider, that people's experiences of digital identity are produced. Such lived experiences occur at points of digital authentication, where user credentials are verified through details stored at the time of registration. Behind the interface is the core principle of digital identity, and the heart of this chapter: authorisation to access provisions is conditional to the successful authentication of the individual. Authentication is, in turn, based on the matching of the person's details with those registered at the time of enrolment. In this chapter, I detail this founding principle, and then review different, but intertwined perspectives from which digital identity has been studied across disciplines.

What's in a Card?

'For the third time, I have not received my ration card'. Aisha is a middle-aged woman living in a slum colony at the periphery of one of the main cities in the Indian state of Kerala. In telling her story, she does not sound angry or vengeful. She is only tired, extenuated by a wait that has protracted for months and that denies her the key document to access the Public Distribution System, India's food security scheme on which millions of households depend for subsistence. The scheme, acronymised as PDS, provides essential goods (primarily rice, wheat, sugar and kerosene) to below-poverty-line (BPL) households at highly subsidised prices. But in July 2010, the government of Kerala registered a backlog of about six lakh (600,000) ration card applications, which resulted in a severe hurdle in access to a vital anti-poverty programme (Masiero, 2012: 4).

Now, in August 2012, Aisha needs her ration card. In the PDS, receiving food rations is conditional to being recognised as an entitled user, in virtue of a targeted system where the ration card – stamped by the ration seller every month at the collection point – determines entitlement. Ration cards in Kerala have different colours according to poverty status: in 2012 they were blue for above-poverty-line (APL) households, pink for BPL and yellow for Antyodaya Anna Yojana (AAY), the poorest of the poor who are entitled to greater quantities of subsidised goods. For the third time Aisha, a slum dweller living below the poverty line, has spent the day queueing by the local Taluk Supply Office, the bureau where ration cards are dispensed, hoping to collect a document that is likely stuck, with many others, in the backlog for which the Kerala Rationing Officer responds.

The PDS is India's largest food security programme. As noted by Mooij (1998), its origins are in the rationing system introduced in colonial Bombay in 1939, at a time of low production of food grains per capita and high reliance on imports. During wartime food grain prices increased sharply, leading to the creation, in 1942, of a Food Department centrally aimed at price control. The Department was tasked with buying food grains from surplus provinces and selling them at below market price to food-deficit ones, seeking to ensure fair redistribution of essential food items across the land (Mooij, 1998).

After the country's independence from Great Britain, the Green Revolution in the early and mid-1960s strongly reduced India's dependency on food imports. The nation's new self-sufficiency in food grain production, still articulated across food-surplus and food-deficit areas, led to the launch, in 1965, of the PDS under the aegis of the newly created Food Corporation of India (FCI). The mandate of the FCI was that of 'enabling the government to undertake trading operations, through which it will influence market prices' (Government of India 1968, cited in Mooij 1998: 84). Under the FCI, the PDS mirrored the old tasks of the Food Department: it procured goods at the central level and redistributed them across states, through a network of shops, known as fair-price or ration shops, through which quotas of rationed goods were collected, in a universal form, by households.

The PDS saw a time of high offtake in 1965–1990, when food grains traded by the FCI increased from 10 to over 18 million tons and the number of ration shops went over 350 thousand (Mooij, 1998: 86). The programme's fiscal burden – with the size of subsidy increasing from 0.04% of GDP in 1970–1971 to 0.5% in 1990–1991 – came, however, under fire during the fiscal crisis suffered by India in the early 1990s (Ahluwalia, 1993). This burden, combined with substantial leakage to non-poor recipients, was defined a 'meagre transfer at exorbitant cost' by the World Bank (Radhakrishna & Subbarao, 1997). In a context of structural adjustment, it was recommended to trim the system down to a programme targeted to BPL residents. In June 1997, the institution of a targeted PDS redesigned the programme, making it targeted to the BPL (with the institution of AAY, for the poorest of the poor, in 2000) and allocating subsidised goods to states based on the relative poverty of each, estimated at the central government level. A targeted PDS was introduced in all Indian states, with the exception of Tamil Nadu, which maintained and still maintains a universal system.

Commentators are divided on the impact of the shift to a targeted PDS. One stream of thought points to how the central government's fiscal expenditure was reduced, resulting into greater ability of the government to cope with a largely poor and malnourished population (Umali-Deininger & Deininger, 2001; Tritah, 2003; Ramaswami & Balakrishnan, 2002). But a large stream of literature points, instead, to the rough consequences that targeting had on PDS users and, relatedly, on the ration shop owners (*ration dealers*) whose customer basis suddenly shrunk (Swaminathan, 2002, 2008; Khera, 2011a, 2011b; Drèze & Khera, 2015). States like Kerala, where my doctoral research took place in 2010–2014, suffered especially heavy impacts from targeting: with a state poverty level estimated at only 25%, many needy families were classed as APL; hence confined to a meagre subsidy effectively approaching market price. Even if the poverty line was then re-estimated at 42% by the Kerala government, which paid subsidy for the difference (Swaminathan, 2002), Kerala witnessed a major crisis for ration dealers (Krishnakumar, 2000; Nair, 2000). The debt incurred led to a wave of ration dealer suicides (Suchitra, 2004), whose impact was still strong at the time of my fieldwork.

Whichever position is taken in the PDS debate, one point stands still. The move to a targeted system – replacing a universal one, which all households could avail – generated a new, massive relevance of identification, and especially of its role as a basis for recipients to claim entitlements. Colour-coded according to poverty status, ration cards in Kerala embodied a core principle: the person's verified identity, enmeshed in the card with poverty status, was the basis for receiving PDS entitlements. Not only was it needed for recognising the person, but also for assigning entitlement, with AAY and BPL entitled to greater quantities and sub-sidies. Paper-based and physically stamped, or otherwise marked, by the ration dealer at the time of collection, India's ration cards epitomise the principle that authorisation to access services requires successful authentication of users.

A Founding Principle

Contextualised in India's long history of documentation practices, the genesis and *social life* (Kopytoff, 1986) of India's ration cards is extensively narrated in historian Tarangini Sriraman's (2018) book 'In pursuit of proof: A History of Identification Documents in India'. Sriraman takes a deep ethnographic approach to narrating the evolving relations between documentation practices and the welfare-based history of the Indian state. Understanding such practices, argues Sriraman, involves tran-scending a biographic approach (Appadurai, 1986) based on ID documents as the artefacts of a sovereign state. It is through a focus on the popular making of ID documents, and on the active role of the enumerated in document production that the history of the many documents shaping India's welfare history is illuminated (Sriraman, 2018: 4–5).

In drawing on Sriraman's ethnographic approach to study ration cards, a core aspect of the document stands out: at the time of my fieldwork, Kerala's ration cards were no digital object. What was digitised was not the card itself, but the Ration Card Management System (RCMS), a workflow management programme instituted by Kerala's Ministry of Food and Civil Supplies in 2010. RCMS was built as an e-governance solution aimed at computerising the main phases of the ration card release process: application by the user; processing by the Taluk Supply Office (TSO); and delivery of the document to the user on TSO premises (Masiero, 2012). Digitising the ration card flow was meant to help process applications more smoothly and effectively, as officers at the Kerala Rationing Collector discussed with me in 2010. While stories of repeated frustration, like Aisha's, were heard frequently at the time, a 'digitisation for development' logic resounded strongly in the voices of the government officials that I interviewed.

The card itself, however, remained a starkly non-digital artefact. Its very archi-tecture, as a barcoded booklet with a set of empty spaces for stamps, was that of a document meant for physical use; the card was stamped by the ration dealer in the fair-price shop at the moment of collection of rations, a monthly occasion for users to collect their quota. As this book will show, several Indian states have now adopted digital or 'smart' versions of ration cards; both the digital versions, to be illustrated later and the physical ones embody the principle that authentication – through user credentials provided at registration – is necessary for authorisation to access a service. In its essence, the ration card materialises the principle on which digital identity architectures are built; for the user to access a given service, the provider needs to know (a) that the user is who they claim to be and (b) that a given user, who claims a certain identity, is entitled to certain provisions in virtue of the identity in point. For example, citizen X whose household is BPL is entitled to 18.75 kilos of rice at one rupee per kg; 16.25 kilos of wheat at two rupees per kg; and two kilos of atta, a type of whole wheat flour, at 12 rupees per kg (these were the Kerala entitlements for BPL in 2011/2012).

Crucial to the ration card is its nature as a household-based document. Moved from an individually issued form in July 1952, the card is, notes again Sriraman (2018: 8), a reflection of the use of 'family as a category of governmentality' as made by the Indian state: it is in virtue of household membership, rather than their own individuality, that people are subjects of rights in the PDS. Such a nature deeply shaped the type of requests made by PDS users through RCMS in Kerala: a household-based document requires addition of members, e.g. in case of a new birth, and deletion, in case of death or indeed of the creation of a new household through marriage. A household-based nature, Sriraman (2018: 8–9) continues, shapes welfare distribution in that an individual cannot, by design, be a member of more than one household at the same time. Coded into a household-based document, PDS entitlements remain embedded in the authorisation-authentication nexus that the ration card epitomises.

As ration cards, their role in the PDS and their seeding with Aadhaar credentials are discussed throughout this book, two caveats are in order. First, a discussion of ration cards needs to be mindful of how the document varies across states. The Kerala case discussed here is particular in its colour coding of cards; for instance Karnataka, one of the states bordering Kerala, presents green or yellow ration cards where entitlements vary with poverty status and the size of households. Several PDS decisions are also taken at the state level, while the central government builds and enforces policy guidelines, essential items such as ration distribution days, the policing of ration shops and guidelines for food inspectors are decided upon at the state level. With my research having taken place in the states of Kerala (2011–2012) and Karnataka (2014 onwards), stories told here draw largely on these states, and are put in a national perspective when discussing issues of design justice in Chapter 6.

Second, since 2017 PDS entitlements and the colour coding of Kerala's ration cards follow the new ration card categories introduced by the National Food Security Act (NFSA), which replaced the distinction of APL and BPL households. A measure augmenting PDS coverage to about two-thirds of the Indian population, the NFSA classifies *Priority Households* (*PHH*) and *Non-Priority Households* (*NPHH*) by criteria that reflect both income and other forms of societal advantage, such as ownership of assets, differently abled household members or the presence of an income taxpayer (Prabhakar, 2017). The new ration cards of Kerala appear in four colours: yellow for the most needy sections of society, a substantial part of whom are AAY beneficiaries; pink for priority households; blue for non-priority households who still receive the state subsidy; and white for general non-priority households. Following the new classification introduced by the NFSA, the new cards of Kerala still embody the same logic: to different status of need, now reflected in the NFSA priority assessment, correspond different entitlements codified in different colours (Manorama, 2017; Mathrubhumi, 2017).

The Architecture of Digital Identity Systems

Digital identity systems, it can be argued, have leveraged technology to embed and crystallise such a long-established logic. A repository of users' demographic (and, more often over time, biometric) data is at the core of a digital identity system's functioning, and as illustrated in Figure 2.1, it is through that repository that data are verified at the time of authentication. *Boundary resources*, a large umbrella term for Application Programming Interfaces (API) and Software Development Kits (SDK), enable the conditionality on which digital identity systems are based: connecting the identity database to external agencies, such resources enable third parties to avail user verification. In this way, service providers can make access to their services conditional to authentication, so that, for example, the person who collects food at the ration shop – be they recognised from a picture, a fingerprint or an iris scan – can prove to be who they claim to be and, on that basis, be authorised (or not) to access entitlements.

Viewed in the light of the authentication-authorisation logic, digital identity systems automatise the same functions that have long been present in social protection systems. The enmeshing, in Kerala's coloured ration cards, of the person's identity with poverty status is reflected in the digital evolution of the system over the last decade. India's anti-poverty system, which was rebuilt from 2009 around the biometric identifier of Aadhaar (a word that means 'foundation' in several Indian languages), has automatised the function that was earlier played by physical verification by the ration dealer. If in the early PDS it was the ration dealer to check a person's identity through their photograph (and often, through the previous knowledge of the person and household), in the biometric PDS such a function is charged to technology, capable of providing a 1 or 0 ('yes or no') value to biometric verification. Equipped with fingerprint readers, Kerala's ration shops do not anymore require the ration dealer to question users: it is the machine that recognises entitlement or not of the user to a certain provision.

Read in this light, digital identity has not brought stark innovation to the world of service provision. Rather than challenging a central principle, that of conditionality of authorisation to authentication of users through credentials, it has

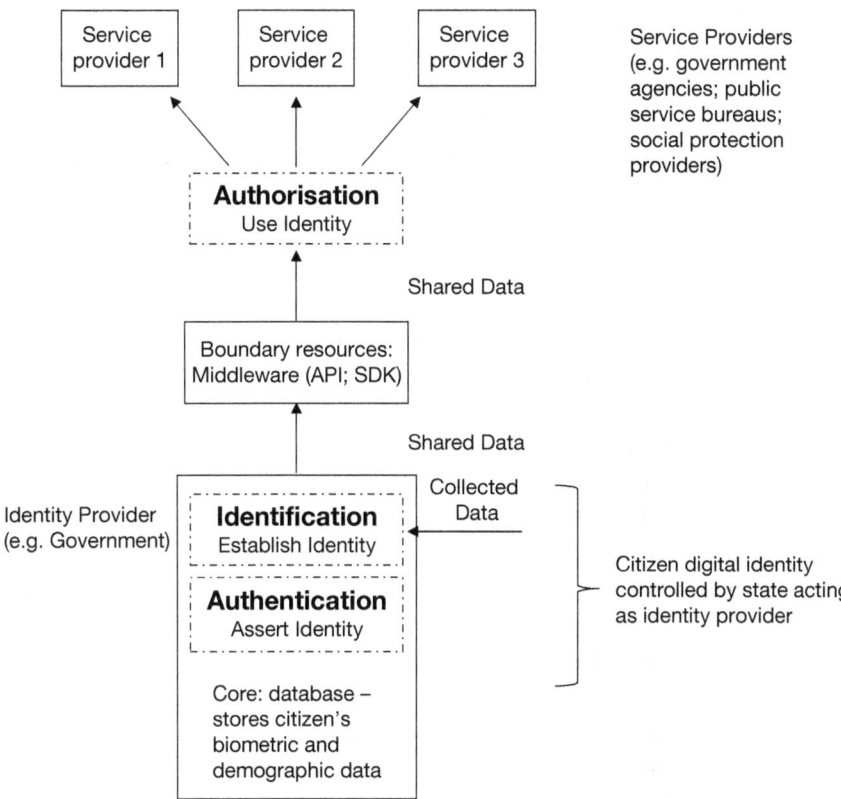

Figure 2.1 The Architecture of Digital Identity Systems
Source: Masiero & Arvidsson (2021: 907) (adapted from Nyst et al., 2016: 12).

merely translated such a principle into technology, curbing access to services to the non-authenticated (arguably classified, by the system, as non-entitled). Rather than being a transformative technology, digital identity seems to encode a very long-standing principle: only those entitled to a given programme should access it, and the right to the same – predicated on the identity-entitlement matching – should be denied to those for whom identity matching fails.

Foundational and Functional Identity: A Firm Distinction?

Two main forms of identifiers are recognised in the digital identity space, illustrated, as in Gelb and Clark (2013), as identity systems of *foundational* and *functional* types. A foundational identity system consists of a centralised architecture built to manage identity information for the general public, and to provide identity proof for a wide variety of public and private services: identity cards, passports and identity documents are all, in principle, classed as foundational. A functional identity system, on the contrary, is designed to meet the needs of an individual sector, and is not designed to be used outside it. Voting cards, driving licences, social security numbers and tax cards are classed as functional, based on their specificity to a given sector of the economy or the government.

India's ration cards pose, however, a question on this dichotomy. On the one hand, the established – and widely used – partition of the two types of identification systems is useful in classifying the different architectures of identification and organising them according to their purpose. It is true, as in Gelb and Clark (2013), that a fundamental difference exists between systems that provide identity proof across sectors – for example national identity cards, or breeder documents for civil registration – and systems whose capability of identification is by design confined to a given field. The dichotomy also helps organising systems according to their practical utilisation, recognising and limiting the domain of action of each.

On the other hand, the conflation of identity and poverty status in the same document, the ration card, raises questions about the definition of borders between documents of foundational and functional types. As Gelb and Clark (2013: 13) note, the ration card is widely used in India as an identification document, more so now that its connection to the Aadhaar database has enabled links to biometric authentication. This leads a functional document – the ration card is built for the PDS system, hence functionally linked to access to a food security scheme – to transcend into a foundational type; people do use their ration card to authenticate at government bureaus, and even for services such as banking and collection of goods from shops outside the rationing system. While useful in classifying different identity systems, the foundational-functional dichotomy has elements of blurriness that narratives in this book will need to take into account.

Digital Identity and Development: Theorising the Link

The story of India's ration cards and of their cruciality for people to access essential entitlements in the national food security system (Sriraman, 2011) opens up a broader discourse on identification for development. An established orthodoxy links identification to human development goals, with legal identity, the core of Sustainable Development Goal (SDG) 16:9, being related to key dimensions of development (Dahan & Gelb, 2015). Such an orthodoxy builds and sustains the epistemological basis for the diffusion of digital identity systems especially in contexts of vulnerability.

A key problem, as we approach questions on the fairness of digital identity systems, is on *how* the link between digital identity and development is articulated. Dahan and Gelb (2015) detail how the legally identified user is made visible to the state and other service providers and can, on that basis, be a subject of rights and entitlements. In this sense, the person's *visibility* is arguably the basis for a route to fairness, with the reverse also being true: it is not accidental if identification schemes are found to gain support especially in societies where anonymity leads to exclusion from services of crucial importance (Murakami Wood & Firmino, 2009). Making each person visible, argues the logic of identity for development, affords considering each person as an independent subject of rights, which is an essential basis for the 'just, peaceful and inclusive societies' that SDG 16 refers to.

As theorised in Masiero and Bailur (2021), at least two dimensions feed into the link between digital identity and development. One pertains to the key *provider* of identity: while this is historically recognised as the state or government, engagement with digital identity systems has spread to supranational actors which have large responsibilities in serving the needy. This applies, in particular, to organisations that require identification of their recipients for the provision of humanitarian aid. Another dimension pertains to the *use* of technology: digital identity is connected to aims that pertain to the interacting domains of access to fundamental services, inclusion of minorities and humanitarian assistance. Both dimensions are represented in Figure 2.2, and the three interlinked uses of technology are examined below.

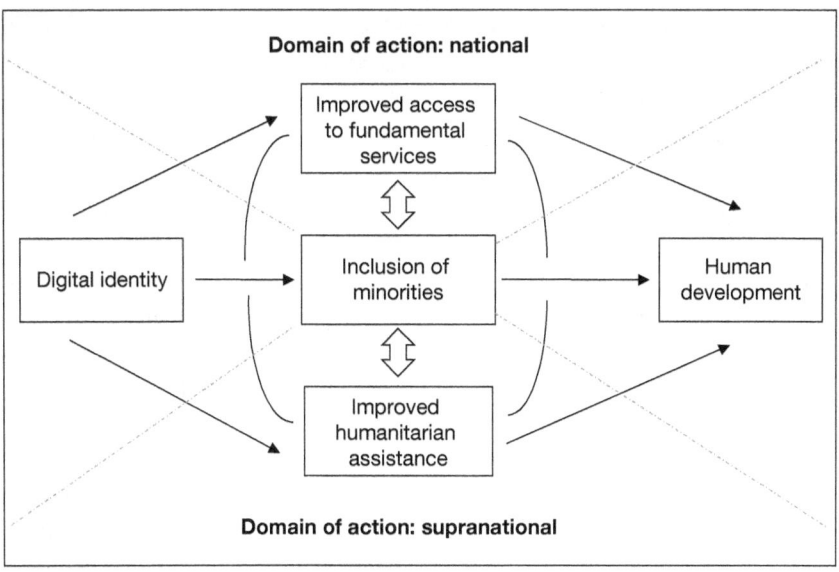

Figure 2.2 Digital Identity and Development: Theorising the Link
Source: Masiero and Bailur (2021: 5).

Improved access to fundamental services: Research has, as noted above, constructed digital identity systems as technologies that afford providing entitlements to individuals. Such entitlements consist, in the first place, of the right to receive government services, provided on the basis of correct identification. An important class of fundamental services are social protection schemes: these are defined, note Devereux & Sabates-Wheeler (2004: 1), as 'all public and private initiatives that provide income or consumption transfers to the poor, protect the vulnerable against livelihood risks and enhance the social status and rights of the marginalised; with the overall objective of reducing the economic and social vulnerability of poor, vulnerable and marginalised groups'. While services such as education and healthcare may have a universal character, identification in targeted social protection schemes acquires additional value due to targeting, meaning that to qualify for the scheme the person must meet particular requirements, for example in terms of poverty status.

Making the person visible, the orthodoxy continues, solves the conundrum of social protection, which is that of ensuring access to all and only those individuals entitled to a certain scheme. As noted earlier, this means solving two problems: an *exclusion error*, meaning the erroneous exclusion of genuinely entitled individuals, and an *inclusion error*, meaning the erroneous inclusion of non-entitled subjects. By matching an individual's identity to their entitlements, digital identity systems propose to solve both problems: exclusion errors are fixed, on the one hand, by the provision of legal identity to those who cannot prove entitlement (for example undocumented or displaced persons); inclusion errors are addressed, simultaneously, by the ability of digital identity to subordinate authorisation to successful authentication of individuals. Through their architecture, digital identity systems enable authorisation of all and only those who are entitled to a certain service; this enables providers to offer services more effectively, positively impacting the quality of users' lives.

Inclusion of minorities: By fixing the core issues of identity recognition, digital identity moves a step beyond the improvement of service provision. With its authorisation-enabling architecture, digital identity promises special support to members of vulnerable groups, who need to prove their status to governments and service providers to receive assistance. Multiple vulnerabilities are produced by identification practices: the contestation of identity, requiring a country's citizens to prove who they are, may put people's right to nationality in peril. Groups whose citizenship is actively contested, such as Kenyan residents of Somali origin (Weitzberg, 2020b) or Haitian-descended people born and living in the Dominican Republic (Hayes de Kalaf, 2019, 2021), similarly risk being rendered stateless by the very same practices that should legitimise their entitlement to nationality. In this landscape digital identity, blending once again with policy (Whitley et al., 2014), promises to include vulnerable groups in the social system, acting as a restorer of rights that other practices have put in danger.

Improved humanitarian assistance: While the provision of legal identity is historically associated to nation states and their sub-divisions, supranational providers are increasingly engaging the digital identity landscape. One reason for this lies in the ability of digital identity to endow the aid recipient, who may struggle to prove their status, with the means to be recognised in the absence of breeder documents (World Bank Group, 2021). Digitising the provision of humanitarian aid is depicted as an empowering mechanism, where identification enables the provision of essential items to households in need. With the increased use of data in the humanitarian sector, the previously state-centred logic embraces the supranational domain, extending the proclaimed benefits of digital identity to subjects made vulnerable by external circumstances (Madon & Schoemaker, 2021).

Access to fundamental services, inclusion of minorities and improved humanitarian assistance are all embedded in an orthodoxy linking digital identity to development. Closely intertwined, these three channels give substance to the promise of a 'better world' (Walsham, 2012) to be achieved with digital ID. Predicated on the visibility that identification affords for the individual, these dimensions articulate the route to fairness that digital identity, targeted to communities in need, promises to achieve.

A Problematic Link

The link between digital identity and development is epitomised in its orthodoxy. But the extent to which, and ways how, the promise of such an orthodoxy is being achieved need to be studied from the perspective of digital identity users. This book takes up this challenge. Going back to the PDS and its use by entitled residents opens a window of observation on such questions.

April 2018. Adeela is a long-term PDS user, whom we meet by her house in the periphery of one of the main cities in the state of Karnataka. We asked her whether, in the place of the subsidised food rations that she receives under the PDS, she would prefer receiving a cash transfer of the same value: our question comes from the fact that a cash transfer policy, aimed at reducing the leakage brought by the PDS, has been advanced as an alternative to the current system. If functioning properly, such a transfer would allow Adeela to buy food grains, sugar and other goods in regular shops, avoiding long queues at the ration shop and the risk to have her households' rations partially, or even completely, taken away and sold to the private market. India's Economic Survey 2015/2016 (Government of India, 2015) proposes a cash transfer policy as a strong alternative to the PDS, devised to do away with distortion and corruption.

Cash transfers – conceived as transfers to bank accounts made for BPL recipients, designed to afford the same protection of in-kind subsidies – were hailed by the central government as a solution to the leakages of the PDS. The case for cash transfers is made by general economic principles: they are built to eliminate economic distortion, meaning the alteration in optimal market conditions caused by the intervention of the state. It is argued, along the same line, that leakages undermine the effectiveness of product subsidies, because incentives to divert goods to the market ultimately lead to scarcity of the goods that remain available to entitled beneficiaries (Government of India, 2015: 52–56). This is why, in conversation with PDS recipients, frustration is common: at the moment of collecting rations, the ration dealer may declare that there are no goods remaining for the month, and instead divert goods on the open market for a direct profit (Khera, 2011a).

Much debate has unfolded on where the main problem of food distribution lies, whether in leakage of the PDS to the non-poor (cf. Bardhan, 2011) or in inaccurate determination of poverty status, resulting in needy households becoming unable to claim benefits (cf. Swaminathan, 2002, 2008). It has, as well, been argued that statistical indicators, rather than conveying 'scientific objectivity', present clear signs of social and political construction (Krishnan, 2023); this is reflected in the very determination of state-level poverty lines, which excluded many needy households from the targeted PDS (Swaminathan, 2002). Access to entitlements, in the form of goods collected in ration shops, is prioritarian for recipients; this is not to underestimate inclusion errors, where non-entitled people are erroneously supplied goods. But it is to say that exclusion errors, where genuinely entitled people do not receive their entitlements, persist under the premises of digital identification, which alone does not involve mechanisms to ensure access to basic-need goods where authentication fails (Khera, 2011c; Masiero, 2015a, 2020).

But would cash transfers, then, not solve it all?

In a chapter programmatically titled 'Wiping Every Tear from Every Eye: the JAM Number Trinity Solution', India's Ministry of Finance sustained that, by replacing the PDS with a proper cash transfer system, the distortion induced by subsidies would be eliminated (Government of India, 2015). Such a cash transfer system would rely on three major pillars: a zero-balance bank account programme for all – the Pradhan Mantri Jan Dhan Yojana, shortened as Jan Dhan Yojana, literally 'Prime Minister's Public Finance Scheme'; Aadhaar as the national digital identification system; and mobile phones, through which transactions can be approved and received. Acronymised as JAM – Jan Dhan Yojana, Aadhaar and mobile phones – the basis of the proposed cash transfer infrastructure would do away with the leakage caused by subsidies, erasing the diversion to the private market that characterised the PDS.

But Adeela's reaction to the cash transfer perspective, when we ask her the question on whether she would prefer this, is a firm negative answer. And strikingly, her response is echoed by the large majority of the PDS users to whom we asked the same question.

Our research in Karnataka (Masiero & Das, 2019) revealed an overarching reason for that. While cash transfers have become an established part of the nation's social protection strategy (Raghavan, 2021), subsidised rations are not always of the expected amount. And can be delayed, diverted, even skip a month or more. But our respondents consistently depicted them as a material reality on their tables. It is food rations that people are used to, and it is the ration that, differently from a money transfer perceived as risky and uncertain, brings essential goods to the table of BPL families (Puri, 2012). Related to this are the intra-familial dynamics of cash transfer distribution, with cash being sometime steered away from food by male family heads, while the NFSA prescribes that 'women of eighteen years of age or above to be head of household for purpose of issue of ration cards' (NFSA, 2013). As a respondent told us in an urban setting in Karnataka, 'at least if we get ration, we have something in the house to eat'.

Not only is Adeela's answer negative, but she is visibly scared by the possibility of a shift that would dismantle the PDS in favour of direct cash transfers. For as much as economic principles can hold in theory, they need to come to terms with the reality lived by people for whom that ration, which brings the secure materiality of food on the table, is the item that counts most. A materiality that a cash transfer system, in a context of uncertain access to banking transactions, is perceived to be capable of dismantling from the basis. In fact, surveys of PDS users have been historically consistent in revealing strong preference for the PDS as compared to a potential move to cash transfers (Aggarwal, 2011; Puri, 2012; Khera, 2014), a topic to which we will return in Chapter 5.

This book illuminates many stories of which Aisha, Adeela and many other digital identity users reveal the traits. These are stories of people who struggle with the effects of digital systems purportedly designed to be fair, and claimed, in fact, to enhance fairness in the lives of vulnerable users. But on the other hand, the users whose stories are collected here are harmed by the same systems. It is the forms of this harm, and of the injustice generating it, that the following chapters illuminate.

Three Views of Digital Identity

The stories told in this book need to be positioned within a literature of reference. Providing such a background is, however, a more complex undertaking than it may appear. In its multiple definitions, digital identity is a highly interdisciplinary topic; a mapping of the digital identity literature spans across fields of research and practice, making the literature difficult to organise.

Forms of digital identity have been the subject of numerous taxonomies. The Identification for Development (ID4D) initiative of the World Bank divides digital identity systems among foundational and functional, also recognising multiple sub-types that different systems present (World Bank Group, 2019). While acknowledging such an undertaking, I take a different one; to map the digital identity field I focus, rather than on the type of technologies, on the *theoretical perspectives* taken on such technologies. By *theoretical perspective* I mean a set of assumptions on what digital identity is and how it works, which influence tenets on digital identity systems and provide the theoretical ground for studying them.

Digital Identity as a Datafier

Sticking to the definition of digital identity used in this book, a *datafier view* sees digital identity as a converter of individuals into machine-readable data. This makes digital identity a mode of *datafication*; what digital architectures datafy is the complex, physical human identity of the user. In the phase of identification, when credentials are registered in a central database, the user's identity is converted into data, creating a digital record that can be administrated through machine read-ability (Srinivasan & Johri, 2013). It is this transition from identity to digital data that underscores datafication, and that turns the individual into a data subject which can be administered (Singh, 2020; Madon & Schoemaker, 2021; Martin & Taylor, 2021a, 2021b; Chaudhuri, 2022).

A datafier perspective calls the question on the rationale for datafication. The identity for development orthodoxy illuminates such a rationale; by combining the identity of individuals with their entitlements, datafication affords combating inclusion and exclusion errors at the same time. The digital identity-entitlement matching is based on machine-readable data, on whose basis service providers can include all entitled users and exclude all non-entitled ones. It is on the basis of the principle of matching users with their entitlements that digital identification technologies are presented, to governments and international organisations, as identity 'solutions' to global challenges, and that a market is created around them (Martin & Taylor, 2021b).

At the same time, a datafier vision is the basis on which the detrimental consequences of datafication are studied, on the same users that digital identity architectures are supposed to benefit. Studying India's social protection system, a stream of research illustrates how technology is participating in the denial of vital assets to individuals and households (Drèze & Khera, 2015, 2017; Drèze et al., 2017; Hundal et al., 2020; Muralidharan et al., 2020). These studies show that when applied to social protection schemes, digital identity systems have produced measurable exclusions: one of the studies argues that such systems can cause 'pain without gain' for the intended recipients, increasing the opportunity for harm without this resulting into measurable benefits (Drèze et al., 2017).

But the problem goes beyond social protection. With the increasing datafication of the humanitarian sector, to be converted into data is not only the identity of state citizens, but that of persons in need of different forms of humanitarian aid and action. While the principle of datafication as an improver of service provision, matching identities and entitlements for people, still holds, so does the presence of issues embedded in the very same mechanism; in particular, datafication can exclude individuals whose data are not correctly matched to the dataset claimed to be the source of their entitlements. In a poignant example, the study of datafied refugees in Jordan, Lebanon and Uganda by Schoemaker et al. (2021) shows that changes in datafied records – in particular changes in the registered presence of a household head – make it possible for households to alter the amounts of goods received, directly linking data representations to the effective social benefits that refugees receive.

In addition, beyond literature on digital ID as a datafier, attention is placed on literature focused on *what people do* as a result of datafied identities. Largely set in the anthropological space, this literature sees important work by Sriraman (2013) on the 'piecemeal pedagogies' engaged by poor women in a slum area of Delhi: lacking access to institutional instructions on the process of obtaining a ration card, women educate each other on the steps for accessing ration cards, relying on common networks and reciprocal assistance. In a similar vein, Shakthi (2020) ethnographically explores the use of Aadhaar to create unique identifications in the National Skills Registry (NSR), India's unique identification project to identify employees. Her work studies the ongoing self-disciplining effects of biometric technology applied in the workplace; beyond datafying people's identity, Aadhaar shapes the behaviour of the datafied, whose work paces and office presence are affected by identification practices, ultimately leading to datafication of the working life.

The Special Section of South Asia: Journal of South Asian Studies, titled 'Aadhaar: Governing India with Biometrics' and edited by Rao and Nair (2019), goes into greater detail on how the codification of identities into India's Aadhaar shapes the behaviour of the datafied. In a paper on the Aadhaar Enabled Bio-metric Attendance System (AEBAS), Solanki (2019) notes how Aadhaar is instrumental to a system that infuses attendance monitoring in the core of India's bureaucracy. The system, Solanki (2019) argues, is partially, and some-what unintentionally, bypassed by members of the public sector: what it dis-plays is however the use that a large bureaucracy makes of datafied identities, plying them to the service of large-scale workplace control. In the same Special Section, Baxi (2019) studies the reception that early announcements of Aadhaar (initially termed the Unique Identification Project, or UID) had among civil society organisations. UID, she argues, met the politics of organisations that had been engaged for a long time in providing identification documents to diverse groups of marginalised people; yet, datafied identities became functional to maintaining exclusions of certain groups from the scheme, crystallising a dynamic of silencing which organisations could not reverse.

Finally, the proposition of algorithmic welfare in terms of Artificial Intelligence (AI) has become prominent in the narrative of AI for development, or more broadly, 'AI for Social Good' (Iazzolino & Stremlau, 2024). Recent studies show, however, the problematicity of datafication when inscribed in algorithms that, tasked with crucial decisions on welfare entitlements, result into further exclusions (Chaudhuri, 2022). As Chapter 8 will explore, algorithmic management is becoming diffuse in social welfare; in a study of an algorithmic system in India's state of Telangana, Tapasya et al. (2024) found significant exclusion errors, leading entitled users to be erroneously classed as non-entitled and denied access to subsidies. Intersecting with notions of algorithmic fairness (Marabelli, 2024), debates on 'AI for social good' leave open questions on the nature of datafication, and especially on its effects on social welfare systems on which people depend for livelihoods.

In sum a vast literature exists on digital identity as a datafier, focusing on how digital identity systems enable benefits but also exclusionary outcomes. Widely diffused across disciplines, this view focuses on the datafication process that digital identity involves, rather than on its outcomes or on the platform features that enable them. As noted below, the few platform-centred works on digital identity systems tend to adopt the perspective of the platform owner, centred on the digital identity orthodoxy but risking, in doing so, to 'ventriloquise for the poor' rather than listening to the poor's voices (Breckenridge, 2019). This makes it important to complement this view with user-centred perspectives, in terms of both digital identity outcomes and architectural features.

Digital Identity as a Mediator of Surveillance

In contrast with a datafication view, a perspective of digital identity as a *mediator of surveillance* is well established in surveillance studies (Lyon, 2001, 2009; Whitley & Hosein, 2010; Bennet & Lyon, 2013; Martin, 2019). Differently from the ID for development vision, this view centres on the surveillance powers linked to digital identity databases, where biometric and demographic data of users are made accessible to database owners (Weitzberg et al., 2021). Digital ID systems, this view sustains, can and should be seen as an integral part of the surveillance society theorised by Lyon (2001). The combination of biometric and demographic data with databases owned by public authorities, for example in cases of refugees or displaced persons, results in surveillance affordances that endanger the user, defying the goal of 'empowerment through digital identity' that the orthodoxy states (Cheesman, 2022a).

From Lyon's (2009) landmark book 'Identifying Citizens: ID Cards as Surveillance', a wide literature has originated on the surveillance affordances of digital identity. In surveillance studies, the racialisation of digital identification technology and its connection to public authority databases with the power to profile rather than assist have been widely remarked (Newell, 2020), again with a focus on issues of algorithmic opacity following the introduction of AI in humanitarian assistance (Coppi et al., 2021). Research remarks the effects of such technologies on vulnerable users' justified reluctance to enrol in digital identity databases, with authorities turning such reluctance against users especially when personal data are traded for access to essential services (Pelizza et al., 2021). The result, studies from the mediated surveillance perspective convene, makes digital identity a tool to police rather than assist, with 'authoritarian surveillance' (Akbari, 2021) not only problematising the empowerment rationale, but creating issues for exactly those users whose greater vulnerability calls for protection. What Iazzolino (2021) calls 'infrastructures of compassionate repression', referring to biometric identity management systems for refugees in Kenya, reflects exactly the inherent conflation of alleged care and effective policing in the same digital tools.

Works on digital identity as mediated surveillance are essential to illustrate the double-edged nature of enrolment in digital identity schemes. This twofold nature persists across the public and private sphere, also affecting labour; in a study of Aadhaar's use by informal workers, Krishna (2021) indicates how Aadhaar verification has become increasingly requested for cab drivers and domestic workers. On the one hand, this is motivated by the narrative of 'financial inclusion' by the job opportunities that secure verification enables, on the other, Krishna (2021) notes the surveillant architecture displayed by the process, which requires verification by workers but leaves them vulnerable to customers and external agents, for whom no verification is required. Krishna's (2021) findings sit in the discourse on the power relations that digital ID entails, which pervades Section 2 of this book.

While vast and extending across disciplines, the literature on digital identity as mediator of surveillance finds a limitation in that it does not, or not yet, openly consider the platform features of digital identity systems. This view looks, however, at the shift of surveillance agency from platform owners to agencies triangulating user data with national or supranational authority profiling. Platform characteristics are arguably underexplored in studies of digital identity as mediated surveillance, leading to a need to greater unpacking of the role that platform features and design properties take in such outcomes.

Digital Identity as a Platform

As noted in this chapter, digital identity systems consist of a core database, a set of boundary resources for agencies to build on the core, and services that use the core's identification capabilities. This is a *platform* architecture as defined by Cusumano et al. (2019); with a common core and complements independent of it, digital identity systems offer the 'technological building blocks' to deliver essential services. At the same time, the platform perspective is taken by only few studies of digital identity, with the literature presenting limited recognition of digital identity systems as platforms. In a strong exemplification of this view, Madon et al. (2022) note that the platform architecture informs the very implementation of population management schemes; in practice, it affords actors to 'orchestrate the functioning of a range of local government and third-party players towards service delivery'.

In the Information Systems (IS) literature, some recognition of such a perspective is however present. Two studies of India's Aadhaar take a platform perspective focusing, respectively, on the ability of the Aadhaar platform to guarantee privacy and security (Mir et al., 2020) and on its scalability for services to the poor (Mukhopadhyay et al., 2019). Both studies are representations of Aadhaar as platform rather than datafier and surveillance: they focus on the role of the database as a core for services to be developed, with attention to the boundary resources that make this possible. While confined to Aadhaar, these studies illuminate platform properties by focusing on the platform's core-complements architecture, illustrating its affordances for service delivery and expansion of service agency to actors outside the government (Mukhopadhyay et al., 2019; Mir et al., 2020).

Nevertheless, both studies take an owner's perspective on the platform, centring research on the platform's features and not including users in the study dataset. The result, especially when studying a platform that mediates access to essential services for millions of Indian poor, is an unfortunate risk to take user perspectives for granted, falling into the issue of 'ventriloquising for the poor' flagged by Breckenridge (2019). While perspectives of datafication and surveillance take the users' view, they miss the platform features that largely shape users' situations. But while the platform perspective takes the design properties of platforms as its core, it seems to overlook the user perspectives which are central to lived experiences of digital identity, and that are the core of this book.

As I write this chapter, two publications have quite recently added flesh to the platform perspective. Whitley and Schoemaker (2022) reflect on the sociopolitical configurations of digital identity principles, arguing that digital identity systems are developed by institutions as part of their pursuit of specific goals. With the notion of sociopolitical configurations, they seek to unveil the underlying worldviews that get codified into the technical features of digital identity platforms; in this light, a platform perspective is translated into the principles that dictate the system's architecture, shaping the way its functioning and actions are organised. If we agree that sociotechnical configurations have important implications for the development and use of digital identity systems, it is especially important to consider the design of digital identity platforms, on which we focus in Chapter 6.

Finally, Schoemaker et al. (2023) authored a piece published in 2023 titled 'Digital Identity and Inclusion: Tracing Technological Transitions'. The paper openly discusses how digital ID intersects with perpetration of surveillance, exclusion and privacy breaches; in doing so, it illuminates the evolution from large ID establishment, referred to as 'Big ID' by the non-profit organisation Access Now, to decentralised systems promising 'empowerment' on the basis of 'decentralisation' (Cheesman, 2022a). Keeping a rights' perspective which highlights the importance of 'recentring rights', the article illuminates a platform architectural history whose ongoing phase is centred on *superapps*, whose data assemblages allow the construction of ready-made, commercially useable profiles of digitally identified people. While the term 'Big ID' will come back in Chapter 7, the architectural history of digital identity platforms opens the way for the data justice framework on which the book is structured.

Summary

This chapter has prepared the ground for the book. We have first met the principle of conditionality of authentication to authorisation on which digital identity stands. We have then elicited and questioned its links to development, before illustrating three core perspectives on digital identity research. It is within the landscape outlined by these perspectives that we introduce the book's lens of data justice.

3

DIGITAL ID: A DATA JUSTICE FRAMEWORK

At least two key lessons emerge from Chapter 2. The first is that digital identity is a contentious object, which is difficult to conceptualise as a homogeneous subject of research. Theoretical perspectives viewing it as a datafier, a platform or a mediator of surveillance interact with each other, placing digital identity systems at the intersection across them. Multiple research fields have engaged digital identity, leaving us with a body of work in which diverse perspectives overlap.

A second lesson is in terms of the impact that digital identity systems have on users' lives. Through the stories of Aisha and Adeela, contextualised in the authorisation-authentication nexus that forms the basis of digital ID, the harm experienced by users as a result of the same systems becomes clear. As such harm is explored across the coming chapters, my narration is centred on its impact on central aspects of people's lives. Aisha's frustration for her inability to access essential food provisions, and Adeela's fear for the advent of a cash system that will take food rations away from her household, are indicators of the damage that digital identity systems can inflict on people, and a methodology centred on the encounter between people and technology – as in Corbridge et al. (2005) – is key to explore the implications of such damage.

This chapter proposes a framework for studying such damage. It begins by outlining the notion of *data justice* as a lens through which to examine the unfairness induced by people's incorporation in digital identity systems. In doing so, it positions digital identity use as a form of *adverse digital incorporation* (Heeks, 2022) by which users are harmed rather than helped. Having done so, this chapter articulates the data justice framework that is used in the book: built out of legal, informational and design-related dimensions, such a framework will guide us in the exploration of injustice that the next section consists of. This chapter ends by placing the framework in the context of data justice research, highlighting its relations with other taxonomies built to navigate the data justice field.

The Need for a Data Justice Lens

Early approaches to data justice powerfully state the need for such a concept, in a world where – in virtue of overarching datafication – new forms of justice need to be imagined. The term *datafication* was introduced in the book *'Big data: A revolution that will transform how we live, work, and think'* (2013), by the Economist journalist Kenneth Cukier and Viktor Mayer-Schönberger, professor of Internet Governance and Regulation at Oxford. The term is meant to indicate both a process and its outcome – that is the process through which existing objects are converted into data, and the reality that results from such a process. The twofold nature of data-fication, argue Cukier and Mayer-Schönberger (2013), is in all it does; it is in the process of converting a network of people into data through Facebook, but also in the datafied network resulting from it. It is in the process of converting thoughts into data through Twitter/X or Mastodon, but also in the datafied thoughts appearing on the same platforms. And so on always maintaining its double-edged nature.

Datafication sets the context for a world in which data justice is needed. In her paper 'What Is Data Justice? The case for connecting digital rights and freedoms globally', Taylor (2017) details the new reality of data-induced visibility that makes data justice essential:

> The increasing availability of digital data reflecting economic and human development, and in particular of 'data fumes' (Thatcher, 2014) – data produced as a by-product of people's use of technological devices and services – is driving a shift in policymaking worldwide from being data informed to being data driven (Kitchin, 2016). These granular data sources which allow researchers to infer people's movements, activities and behaviour have ethical, political and practical implications for the way people are seen and treated by the state and by the private sector (and, importantly, by both acting in combination). (Taylor, 2017: 1)

The 'data revolution' (United Nations, 2016) within which Taylor positions the concept of data justice is pervasive, and as of her quote above, it results in levels of data availability that were unimaginable even in a recent past. As she continues to note data produced involuntarily – the 'data fumes' referred to by Thatcher (2014) – constitute a substantial part of such new data availability. On the one hand, a data-for-development logic sees data as a route for more accurate visualisation of matters of interest to development interventions and policy worldwide. In this fashion information and communication technologies, or ICTs, are increasingly linked to the pursuit of development objectives of local and global relevance, first and foremost the SDGs (Heeks, 2022).

But at the same time, the new datafied world requires an idea of justice to be updated and redesigned. The idea of justice, Taylor continues, leads to establishing the rule of law; in the same way, an idea of data justice operates in a datafied space, working to determine 'ethical paths in a datafied world' (Taylor, 2017: 1). In virtue of its ethical mission, data justice is defined in Taylor (2017) as 'fairness in the way people are made visible, represented and treated as a result of their production of digital data'. Visualisation, representation and treatment of human beings are data-shaped, and require a concept dedicated specifically to assessing the fairness with which they happen.

Early understandings of data justice have already explored multiple sides of it. In their article 'Data justice for development: What would it mean?', University of Manchester scholars Richard Heeks and Jaco Renken (2018) offer a taxonomy of data justice that illuminates different, complementary ways in which it operates across people's lives. The taxonomy they propose consists of four dimensions, respectively:

- Instrumental data justice: An instrumental dimension focuses on the fair use of data, zooming on the *outcomes* of the data use process;
- Procedural data justice: In a procedural dimension, the focus is on the extent to which data are handled fairly, and the ways through which such fairness is pursued or hindered. The focus is hence on the *process* through which data are managed;
- Distributive, or rights-based data justice: In this dimension, what counts is the fair *distribution* of data, for example within the organisations that are due to manage them;
- Structural data justice: Built on the shortcomings of the other three, a structural approach to data justice views data as embedded in the *power relations* that characterise a given context.

More discussion is made of a structural approach to data justice. For its direct engagement of power relations, it is this approach that Heeks and Renken (2018) see as most suited to investigate data justice in the so-called developing world, where exogenous forms of injustice (for example unequal resource distributions) intersect with those produced by data. In taxonomising data justice, Heeks and Renken bring a twofold message: first, the mirror image of justice – the injustices that data can produce – can be equally classified and take different shapes; second, a structural dimension is not an addition to the taxonomy – it is instead a transversal dimension that encompasses all others, positioning data injustice in existing contexts of violence and inequality. The study of data justice, they note, cannot be isolated from the systemic forms of injustice already present before datafication.

The new datafied reality, overlapping with injustices existing from before it, illuminates the argument made by Dencik et al. (2022) in their recent book 'Data Justice'. The book opens with the statement that data justice is, first and foremost, a lens; that is a perspective through which to study the datafied reality that we are faced with on a daily basis. Datafication affords new injustices in such a reality, and data justice arises as a paradigm to make sense of them and of how they are produced. It is this lens that enables our exploration of the injustice perpetrated through digital identity systems.

Data Justice and Digital Identity

The rationale behind the fit between a data justice lens and digital identity systems requires greater explanation. What makes such a fit less than immediately understandable is that on the one hand, the digital identity literature presents only few contributions inspired by data justice. While a recent Special Issue of the journal Information Technology for Development featured three such contributions (Martin & Taylor, 2021a; Krishna, 2021; Schoemaker et al., 2021), and such fit has been widely discussed at the Data Justice Conference in Cardiff in June 2023, the turn to data justice of digital identity research is recent, and not inspired by a shared conception of the data justice notion. At the same time, only a minimal part of the data justice literature studies digital identity. Indicatively, Dencik et al. (2022) touch upon digital identity in only two chapters of their Data Justice book.

There are three main reasons grounding the suitability of data justice as a lens to read digital identity stories. By *digital identity stories* I mean the lived experiences of digital identity systems that, such as those from Aisha and Adeela narrated in the previous chapter, populate this book. A first reason lies in the very nature of digital identity systems, which are, as illustrated above, themselves *datafiers* of human identities. The fit of the lens stems directly from how these systems produce digital data, which are the core object of data justice in Taylor's (2017) definition. With the conversion of human identities into data, digital identity systems make people visible, represent them, and enable providers to treat them in more or less fair ways. It is this fairness, and the ways it is hampered, that a data justice lens affords investigating.

But two more reasons support a data justice lens. As noted in the opening, this book is centred on the user as a subject experiencing the act of being digitally identified and intervened upon. The book's core question, on *what happens to the user* as a result of being incorporated in digital identity systems, requires a lens that enables investigation of the lived experience of the user in action. With its focus on the fairness, or absence of the same, in visualisation, representation and treatment of people through digital data, a data justice lens offers a vocabulary to articulate such experiences; it is through a data justice view that stories of fairness, and of people's experience of its absence, are narrated. As the framework proposed below will illustrate, a data justice lens puts forward a conceptual apparatus to narrate such experiences.

A third reason lies in the meaning of *fairness* discussed in the book's introduction. If we use the Oxford dictionary's definition as 'impartial and just treatment and behaviour, without favouritism and discrimination', fairness has both an absolute meaning and a relational one. In absolute terms, fairness results in the experience of just systems that enable user access to essential goods or services, without generating unjust dynamics such as the denial of a ration card for Aisha or the transition to a cash transfer system for Adeela. In relational terms, a fair digital identity system will not result in outcomes that privilege or exclude particular users, hence opposing favouritism and discrimination. Both aspects, absolute and relative, are crucial to the user, which makes a fairness-centred lens such as data justice even more suited to study digital identity.

All these aspects make a data justice perspective a key tool for investigating people's experience of digital identity systems. Importantly, all three elements of data justice – visualisation, representation and treatment of individuals through digital data – are reflected in systems of this type. Digital identity systems *visualise* people, as the conversion of identity into data results into records – then matched with people's entitlements – through which individuals can be seen. Based on this act, these systems *represent* people through their biometric and demographic features, as well as the data collected on them to assign entitlements or provide services. Finally, treatment – the action taken on entitlements, services provided or other consequences of identification – is also directly enabled by representation. The ability of data justice to encompass all three aspects affirms this as a lens on which a framework for the study of fairness in ID can be built.

The experience of unfairness, resulting into harm in core aspects of life as it is for Aisha and Adeela, plays a key part in determining the need for a framework through which unfairness is investigated. A broader categorisation identifies digital identity use as a route to *adverse digital incorporation*; this is defined, with Heeks (2022), as 'inclusion in a digital system that enables a more-advantaged group to extract disproportionate value from the work or resources of another, less-advantaged group'. With a long history investigating the 'digital divide' between the haves and have-nots of ICTs (Warschauer, 2004), ICT for Development (ICT4D) literature has started only recently to engage the ways harm is perpetrated through the same technologies, and the unequal distribution of such harm (Heeks, 2022). Given the pervasiveness of such harm in the so-called developing nations (Lowe, 2022), adverse digital incorporation has fast become an important route to conceptualise the detrimental effects of the digital technologies studied in ICT4D.

This is why, when studying digital identity systems, adverse digital incorporation offers an important interpretive device. Neither Aisha nor Adeela experienced direct negative effects of the non-digital version of the PDS, which had been running for many years. In the case of Aisha, the digitisation of the ration card management process resulted in a backlog that left her application in the pipeline for months, stuck where is still was at the time of my interview with her in 2012. Harm here results from the very material, and urgent, problem of being unable to access food supplies through the PDS. Chapter 4 will explore the most extreme consequences of this issue: research conducted in India's Jharkhand points to hunger deaths from denial of food supplies (Singh, 2019), through the same system that Aisha is hoping to access again. Defined in my framework as *legal injustice*, the situation is one of legal rights and entitlements made conditional to digital identification and authentication.

In Adeela's case, the problem lies in the ability of digital identity to enable a shift towards a system of whose advent she is scared. Two forms of injustice are conflated in her case. First, she does not know that the Aadhaar technology is functional to the transition to a cash transfer system, a transition that worries her due to the impending loss of the food rations on which her household depends. Below I will refer to her situation as *informational injustice*, a state in which datafication determines changes that users are affected by, but are not or not completely informed about. Second, the system is designed in such a way to determine a shift to an architecture, that of cash transfers, that is suboptimal for her. I will refer to this below as *design-related injustice*, a term referring to a situation in which technology design is not fair to users, and indeed causes harm on them.

Both Aisha and Adeela experience harsh effects of being incorporated into digital systems that replace the previous, non-digital PDS. Their stories are key to illuminate digital identity use as a case of adverse digital incorporation, where the user pays the price of being part of the digital system. Their stories reveal legal, informational and design-related aspects of data injustice; it is to these aspects, in their role as constituents of our data justice framework, that we now turn.

Birth of a Framework

Having explored the fit of a data justice lens to study digital identity systems, I will now introduce the theoretical framework that results from such a lens. To do so, the framework should be placed within the research from which its first version, published in Masiero and Das (2019), was developed. The research in point was centred on India's PDS in the state of Karnataka, one of those states in which the incorporation of digital identity in social protection was earliest in time. Having run an independent pre-Aadhaar programme, which recorded the biometric data of PDS users (Raghunandan, 2013), the state of Karnataka was well positioned for the transition of the PDS to Aadhaar-enabled verification, which occurred between 2015–2016 and the present times.

As noted in Chapter 1, Aadhaar is the largest biometric identification programme in the world. Based on 'free and voluntary' capture of ten fingerprints and iris scan for Indian residents, Aadhaar registers each enrollee by giving them a 12-digit, unique number useable for authentication through all public offices in the nation. Articulated as an e-governance infrastructure (Nilekani & Shah, 2016), Aadhaar arose with time as the model for foundational digital identity systems on a global scale. As we get to know Aadhaar through the next chapters, it is important to learn its origins as a central government scheme launched in 2009, to provide a unique identity to all Indian residents for the purpose of better access to public services (UIDAI, 2019).

As Adeela's story reveals, the Aadhaar infrastructure was progressively incorporated in Karnataka's PDS. All PDS users, whose biometric data had already been captured at the state level (Raghunandan, 2013), were demanded to enrol in Aadhaar, through the many enrolment centres disseminated across the state. A free procedure capturing biometric data points, Aadhaar enrolment in Karnataka followed the large popularity that the same process had through political push at the national level. A flagship project of the United Progressive Alliance (UPA) government in 2009, of which Aadhaar's founder Nandan Nilekani was an exponent, the programme was then taken up by the two National Democratic Alliance (NDA) governments that followed in 2014 and 2019, and made into a symbol of digital governance that spurred enrolment rates through the country.

Our research centred on Aadhaar's incorporation in the PDS. Made into an essential requirement for accessing rations, Aadhaar records were integrated with the PDS entitlements recorded for each user, making it possible for recipients to authenticate through their biometric credentials at the ration shop. As we witnessed in our fieldwork, ration dealers in Karnataka now subordinate the provision of rations to successful user authentication with Aadhaar-recorded fingerprints, performed through a fingerprint reading point-of-sale machine located in each ration shop. So organised, the system is built to combat the 'burgeoning rates of leakage' that affected it in pre-Aadhaar times (Gulati & Saini, 2015); with food prices being much lower in the PDS as compared to the free market, a strong incentive is present to divert commodities on the market, where substantial profit can be generated.

As noted above, it is people's lived experience of digital identity that our work enquired. And in this research, based on years of past work on the PDS, our time was mostly spent in ration shops trying to grasp users' experience of collecting rations through the Aadhaar-based PDS. We relied on users' familiarity with the previous state-level biometric system, which I had researched before (Masiero & Prakash, 2015, 2019; Prakash & Masiero, 2015). With the advent of Aadhaar, our lens was squarely placed on how users experienced the Aadhaar-based mode of collecting rations, which constitute substantial components of livelihood for households below the poverty line.

The data justice framework proposed in this book originated during our research. While at the time we were not well-versed with a data justice lens, over time we found strong resonance of a data justice perspective with the forms of unfairness that our research uncovered. As time went on, such forms of unfairness became more neatly inscribed in the types of data injustice that our framework, developed from the ground up, resulted in defining. What this book presents is an evolved formulation of the data justice framework that first appeared in Masiero and Das (2019), enriched beyond our research through the insights of the digital identity literature examined in Chapter 2.

Digital Identity: A Data Justice Framework

Many ways can be devised to read digital identity through a data justice lens. The way used in this book starts directly from people's narratives. It is people's stories of unfairness, as lived through their interaction with the digital identity system, that inspired our construction of the framework and my subsequent refining of it. User stories led us to the derivation of three dimensions of data injustice, respectively categorised in legal, informational and design-related terms, as illustrated in Figure 3.1. Subsequent work through digital identity literature inspired my further refining of these categories, progressively leading to the specification of the framework proposed here.

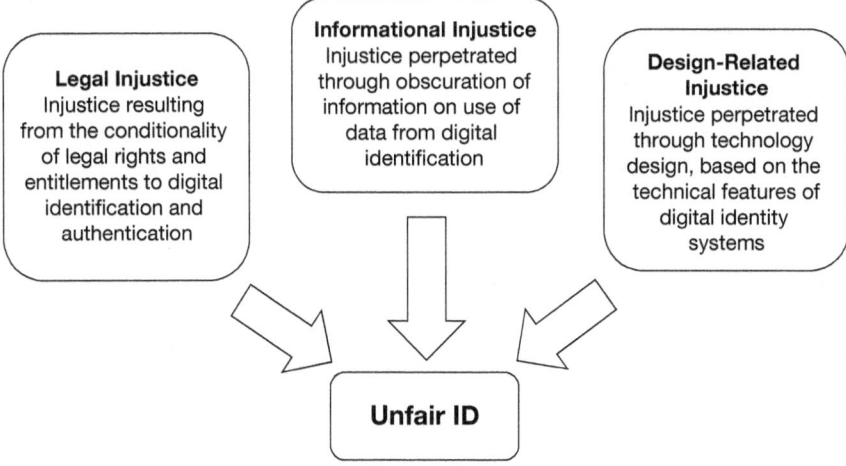

Figure 3.1 Understanding Unfair ID: A Data Justice Framework

Legal Injustice

The first dimension in our data justice framework is *legal injustice*. With the term legal injustice, I mean *injustice resulting from the conditionality of legal rights and entitlements to digital identification and authentication*. A legal injustice is committed when a universal right – such as the right to food, shelter, protection and, indeed, life – is denied by means of a digital identity system. The concept of legal injustice is inspired by the stories of digitally induced exclusion encountered during my research on the PDS, starting from my Kerala fieldwork back in 2011–2012, and has been later corroborated by parallel instantiations of denial of fundamental rights on a global scale.

My first encounters with legal injustice occurred in Kerala in late 2011. At this time, the digitisation of identity for PDS users was in an earlier phase of its history, characterised by the electronic emission of ration cards through the Ration Card Management System (RCMS) described in Chapter 2. Such a phase was markedly different from biometric capture of Aadhaar records, and, yet, it was characterised by the massive backlog of ration card applications due to which Aisha's request was stuck in the pipeline for many months. The result of such a pipeline, which was palpable in Aisha's frustration, was the denial of ration cards to a substantial part of the entitled population: that is, all those applicants whose long wait denied them access to the system. Back then the state of Kerala registered a backlog of over 600 thousand ration cards (Masiero, 2012: 4); for such applicants, the universal right to food was made conditional to a missing document, resulting into legal injustice.

A widely researched form of what is termed here as legal injustice are cases in which entitled users of public services, social protection or humanitarian schemes are denied access to them. While, in Aisha's case, the problem lied with delivery of her ration card, Section 2 of this book will explore the biometric version of such exclusionary dynamics. This happens when biometric point-of-sale machines do not recognise the user as entitled, preventing them from accessing food provisions. Econometric work throughout India widely documents such exclusions (Drèze et al., 2017; Khera, 2017; Muralidharan et al., 2020). Such findings square with our research in Karnataka, where witnessing families queueing together to the ration shop – in the hope for at least one member to be able to authenticate – inspired our concept of legal injustice.

The concept of legal injustice inspires the broader theorisation of digital identity that Chapter 4 details. On the one hand, such a concept emerged in the local reality of India's ration shops; but on the other, it works to understand a form of exclusion on which diffuse digital identity research exists. This refers to all research on exclusions connected to the translation of the authorisation-authentication nexus of digital identity into reality. While inclusion errors are mostly effectively battled through such a nexus (Muralidharan et al., 2020), the nexus alone is not capable of assisting users who are erroneously seen as non-entitled, and are therefore left unserved. Hunger deaths allegedly associated to Aadhaar-based authentication in the PDS (Singh, 2019) illustrate the limit consequences that legal injustice can yield through the undue restriction of legal rights and entitlements.

Finally, research that intersects digital ID with statehood, citizenship and its determination further corroborates the relevance of legal injustice as an epistemic tool. Chapter 4 will deepen such cases; in Kenya, the issue of *double registration* of people appearing in the UNHCR register of refugees has denied the right to a national ID card to large communities, especially of Somali origin (Weitzberg, 2020b). A case of a similar nature is presented by Uganda, where digital identity has resulted into large-scale exclusions of women and elderly people from healthcare and social protection programmes (Centre for Human Rights and Global Justice, 2021). With its focus on the conditionality of rights and entitlements to digital identification and authentication, legal injustice unpacks the exclusionary effects of technologies originally made to improve service provision.

Informational Injustice

Beyond the discussion on access to entitlements, and of how this is systematically hampered by legal injustice, an informational dimension of injustice also plays a role in the framework. By *informational injustice,* I refer to *injustice perpetrated through obscuration of information on use of data from digital identification.* On the one hand, not all users of digital identity schemes are in the position to demand clarity on how their data are handled – that is on procedural data justice in the terms of Heeks and Renken (2018). An example of users who are not in the position to demand such clarity is found again in India's PDS, whose users, once the programme has shifted to Aadhaar, need an Aadhaar identifier to collect their ration, whatever handling is made of the biometric data they give out.

Similar to legal injustice, informational injustice is also a concept that came out of our interactions with PDS users. While conducting research in Karnataka, we were struck by the blank look that many respondents gave us, when asking them how the government will manage or use their data. The reason – we came to understand in conversation – is that the alternative to giving out their biometrics is not receiving PDS rations at all, as well as many other entitlements provided through government schemes. The perceived urgency of registering with Aadhaar led users to do so without questions, in some cases even conflating Aadhaar with other schemes. When asked how her registration process took place, a PDS user told us 'my husband did it for me' (a reconstruction that is surprising, given that Aadhaar registration requires the physical presence of the individual).

But there are deeper aspects of informational injustice. As we first developed the concept, we had in mind the opacity with which the state is known to manage user data. A powerful conceptualisation of such opacity comes from the work of Bidisha Chaudhuri, Assistant Professor of Government, Information Cultures and Digital Citizenship at the University of Amsterdam. Chaudhuri (2021) argues that users of anti-poverty programmes see the state as 'distant, opaque and seamful' through Aadhaar. During our fieldwork, however, we found that the issue of incomplete information is similarly harmful, though in different ways, to outright opacity. A common example was that of Adeela's situation: embracing biometric identification in the PDS, she was not aware of the *grand design* of Aadhaar as a route to move from the PDS to a scheme based on cash transfers. Like other users, she was highly concerned about a possible shift to cash, but was in no position to access information on the functionality of Aadhaar to determine exactly that shift.

Similar to legal injustice, informational injustice is a conceptual device found in fieldwork, but with broad relevance to study unfairness in digital identity systems. The chapter ventures into the notion of information-erasing artefacts; through the work of Cheesman (2022a, 2022b, 2022c), it explores a blockchain-based project providing digital wallets to women refugees enrolled in a cash-for-work programme in Jordan. On the one hand, a blockchain-based architecture was built with the promise of guaranteeing a self-sovereign identity to refugees, maximising financial independence for women otherwise constrained in familial and aid-based systems of sustenance. On the other, the blockchain-based project involved a biometric recognition architecture which produced receipts for withdrawing salary money; such receipts were, however, omissive of key information on worked days, salary earned and time of earning, which left workers in confused and difficult positions. Cheesman's work is leveraged to illuminate the notion of information-erasing artefacts, through which the chapter's story proceeds.

In addition, subsidy schemes developed during the COVID-19 pandemic offer important instantiations of how this type of informational injustice plays out. In Colombia's Ingreso Solidario (Solidarity Income) scheme, data collected from different databases managed by the government were cross-referenced to identify households in need for subsidies (UNDP, 2020: 27). To do so, information was combined from existing databases, charging an algorithm with the decision on which households were to receive the subsidy. But in the machine-led process of entitlement assignation, what remained obscure was how information was combined leaving households uncertain on their subsidy status and on the data that concurred to determine it (López, 2021a).

These cases are used, in Chapter 5, to exemplify the enactment of informational injustice across aid and social protection schemes in which digital identity systems have been incorporated. Both cases reflect Chaudhuri's (2021) notion of the state being seen as 'distant, opaque and seamful' through digital identity use, respectively, in humanitarianism and social protection. In Jordan, refugees are faced with an artefact that, in contrast to previously existing cash envelopes, omits key information on their salaries. In Colombia, citizens are not informed of the algorithmic reasoning behind the decision for them to be assigned COVID-19 subsidies or not. In both cases, information on how data within existing datasets are used and combined is missing, leading a situation of distance, opacity and seamfulness to grow even stronger. Cases point at diverse forms of informational injustice, in which provision of incomplete pieces of information seems as harmful as the absence of information at all.

Design-Related Injustice

The design-related dimension of data injustice is perhaps the one that has evolved most from the original meaning assigned to it in Masiero and Das (2019). By *design-related injustice,* I mean *injustice perpetrated through technology design, based on the technical features of digital identity systems.* Such a definition is inspired by the work of MIT researcher Sasha Costanza-Chock (2020), which shows how injustice can be embedded into technology design and result in direct harm on people. Her work, epitomised by her illustration on body scanners in airports – highlighting how bodies of a non-conform gender identity suffer injustice through technology design – illuminates both the agency of technology in perpetrating injustice and the harm that it causes on technology users.

When we first developed it, the concept of design-related injustice was in much less encompassing terms than it is in its current formulation. In Masiero and Das (2019), the concept was inspired by the narratives of those PDS users who, while trying many times to authenticate through Aadhaar in ration shops, did not manage to do so, hence becoming unable to collect their rations. While this is a situation of legal injustice as we defined it, it also reflects a technology design that is not meant to align with users' main need, which is that of collecting rations rather than seeing such rations denied to people who are not entitled to them. Designed to combat inclusion errors, but at the same time magnifying the exclusion errors caused by the system, the Aadhaar-based PDS had led us to theorise design-related injustice as the injustice resulting from misalignment of the technology with user needs.

The book *Design Justice: Community-Led Practices to Build the Worlds We Need* by Sasha Costanza-Chock (2020), however, inspired a full rethinking of our concept. The core argument of the book is how injustice can be directly designed into technology, as it is in the case of body scanners designed to flag non-conform gender bodies. Inbuilt into technology design, this form of injustice is a lot stronger that the bare misalignment with user needs that we had initially theorised into it. Misalignment still plays a role, as it does in the Aadhaar case where technology prioritises inclusion to exclusion errors, but Chapter 6 shows instances of injustice that result into outright harm rather than simple misalignment. The discussion refers to research on biometric border control and on a large humanitarian-tech partnership in food distribution.

I draw on the work of Annalisa Pelizza, professor of Science and Technology Studies at the University of Bologna, to study design injustice in biometric border control. Pelizza (2020) illustrates how a shift in Eurodac, a system which univocally identifies asylum seekers in European countries through their fingerprints, in 2015 made the Eurodac database interoperable with national police authority databases across Europe. While almost unnoticed by the public opinion, the shift had severe effects for whoever is identified as an asylum seeker in the EU, automatically making their data accessible and useable by police authorities. In Eurodac, interoperability combines the design injustice of undue data sharing with the harm resulting on asylum seekers by enhanced surveillance.

But questions of design justice also depend on *who* designs the technology. And the chapter moves into exploration of a large humanitarian–private partnership, announced in 2019 between the WFP and a technology giant, Palantir, for supporting Optimus, the WFP's supply chain management system. The logic of the partnership is that of relying on Palantir's data analytics capabilities to induce efficiency and cost savings into the WFP's operations, changes that can be life-saving for the needy beneficiaries the WFP caters to. But behind involvement in humanitarianism, Palantir has a history of involvement in practices that Amnesty International (2020) reports as endangering for human rights; commercialised to security agencies of the US government, Palantir's software has been implicated in asylum seekers and migrants divided from family members and subjected to workplace raids, deportation and detention (Hvistendahl, 2021). Keeping in mind Martin's (2023) point on *aidwashing* the reputation of surveillance companies through humanitarian involvement, this chapter interrogates the design justice implications of a partnership that can feed Palantir wide datasets on global food security.

My exploration of design-related injustice relies on the notion of *dark matter* of digital ID, a notion which I counterpose to the narrative on the 'dark side' of IT. Particularly diffused in the field of Information Systems (IS), research on the 'dark side' conceptualises the harm derived from IT as a set of 'side' effects, whose peripheral nature can be combated with incremental improvements in the artefact. A *dark matter* vision draws on Costanza-Chock's (2020) notion of design justice to note how injustice, rather than peripheral, can be inscribed in the very body of the technology; ration shop technologies excluding entitled users, asylum seeker databases interoperable with police authorities and supply chain management systems powered by companies enmeshed with aggressive surveillance are not 'incidentally' unjust, but programmatically scripted to be so. This chapter concludes with the use of Iazzolino's (2021) notion of *infrastructures of compassionate repression*, which portrays the infrastructural embodiment of the conflation of care and policing.

Summary

Born within the study of one programme, our framework has been extended to become a theoretically informed tool to study unfairness in digital ID. Identifying three dimensions of injustice, the framework relates them to the features of digital identity systems through which they are produced. The relation between injustice and digital identity system features is the basis of the core argument of this book that there are forms of injustice – of legal, informational and design-related types – that are produced *specifically* with digital identity systems, and that could not be produced without them. As we venture into Section 2, which has a dedicated chapter for each of the three dimensions of injustice, the relation between system features and unfair ID outcomes will be explored in detail.

As noted above, other taxonomies operate in data justice research. Heeks and Renken (2018) have also put forward a data justice framework, which finds a practical application in the work of Heeks and Shekhar (2019). Studying four mapping initiatives in four cities of the Global South, Heeks and Shekhar illustrate the meaning of procedural, operational, distributive, rights-based and structural data justice, offering practical instantiations of what each dimension effectively means for citizens being 'mapped' through the initiatives. Adding to research on datafication of urban spaces, this research resonates with ours in that it offers a classification of data injustices, as each dimension can be read through its potential to produce justice as well as through the unfairness that can be produced within it.

Unlike Heeks and Shekhar (2019), however, our framework is inspired by the empirical phenomenon of the global diffusion of digital identity systems, whose effects on people needs investigation through a data justice lens. It was this empirical need that led us to the first formulation of our framework, and that inspired my further work to build its transition from the study of one digital identity system to a generalisable theoretical device. As noted in Chapter 2, the authorisation-authentication nexus is the founding principle of digital ID, and it is on this principle that the injustices examined in this chapter are predicated. The next section will show that architectural features of digital identity systems are closely related with the unfairness of ID; making such a relation explicit through the three dimensions, our framework will guide the journey across injustices that unfolds through the book.

Another, more implicit taxonomy of data injustices is proposed by the 'Data Justice' book recently published by Dencik et al. (2022). The book consists of eight chapters, each of which engages an aspect of data-induced change, ranging from a first chapter on 'data and capitalism' to a final, programmatic chapter on 'data and social justice'. Exploring among others the topics of de-westernisation, harm and governance in relation to data, the authors put data justice forward as a lens to study the multifaceted impacts of datafication across human lives. As they state in the book's introduction:

> The privileging of social justice in our analysis of datafication is rooted in an understanding of technology as embedded within an amalgamation of different actors and social forces, and a particular political economy. In this sense, the way data is generated and collected, what it is used for and how, are not inevitabilities but are rather bound up with certain structures, interests and ideas. An orientation towards justice in analysing the intersection of datafication and society asserts from the outset, therefore, that the nature of this intersection is never fixed. It is subject to continuous conflict and negotiation, advanced on the basis of competing visions. (Dencik et al., 2022: 2)

The programmatic intent of leveraging data justice as a lens to study phenomena that directly affect people's lives is well expressed in the Data Justice book. It is with the same intent that our Unfair ID framework is built, using data justice as the theoretical basis to illuminate diverse, humanly relevant dimensions of ID unfairness. The book sits in the data justice literature as a representation of the epistemic role of data justice, and at the same time as a route to explore possible ways to build forms of 'fair ID' by learning from injustice. Our data justice framework, articulated in its legal, informational and design-related dimensions, will guide us through the book's journey from digital ID injustice to resistance.

PART 2
INJUSTICE

4
LEGAL INJUSTICE

Here begins our journey through the injustices of digital identity. It is a journey of multiple stops, each of which tells the stories of people that, interfacing in diverse ways with digital identity systems, have been deeply affected by them. Our first stop is outside a ration shop in a large city of Karnataka, where, on a ration distribution day, people stand in line to collect their monthly allocated goods.

A Family Matter

April 2018. A young woman, Deva, stands in the ration shop line with her mother Ayanka. When their turn comes, Ayanka goes through the Aadhaar-Based Biometric Authentication (ABBA) system implemented in the shop. The ration dealer checks her ration card, inputting her ration card number in a state-provided laptop. He then prompts her to enter her fingerprint in a biometric machine reader to retrieve her identity and entitlement. Differently from earlier versions of the PDS, ABBA requires not only the ration card, but biometric authentication along with it, which in Karnataka is performed with a fingerprint reader separate from the shop's point-of-sale machine.

Yet, Ayanka's fingerprint is not recognised. Instead of letting her through, the system cannot proceed; authentication, the step in which a person's credentials are verified, does not happen for her. All other requirements to access her ration are into place – her ration card is recognised as valid, and having linked the card to the Aadhaar database, her data are present in the system. And still the most crucial phase, the biometric check that should afford her access to the PDS, does not work, potentially leaving her without a ration. This is why Deva is here; with her credentials linked to the same household ration card as Ayanka, her fingerprint is recognised allowing them to receive the ration. 'It never works with her', she says, referring to Ayanka and her attempt to authenticate.

Deva and Ayanka are no isolated case. In our time in the ration shops in 2018, we encountered several cases of smaller and larger family groups lining up together to ensure ration collection. For Ayanka, Deva continues, nobody has been able to explain why fingerprint authentication fails. At the time of introduction of ABBA, all PDS users were requested to link their ration card to their Aadhaar records, a process often performed at the local *seva kendra* (a shop providing a variety of services related to official documents). This process, called *Aadhaar seeding*, provides an extra layer of security that the person effectively is who they claim to be, and, crucially, that they are entitled to the PDS. Behind such authentication policy is the national government's intended fight to leakage from the PDS; after many years of what the Justice Wadhwa Committee (2010) referred to as 'rice mafia', meaning diversion of PDS goods to the private market, biometric technology became the promised guarantor that the person collecting the ration is who they claim to be, and, crucially, is entitled to ration collection.

Two questions arise at this point. The first one is on whether a 'last mile' system – one implemented, such as ABBA, at the last mile of the supply chain, where the person encounters the ration dealer – is well suited for combating the illicit diversion that has plagued the PDS for many years. Artefacts, noted Winner (1980), are unescapably political; more specifically, they 'have politics' that they carry as they perform their functions, and that inform how these functions are enacted. In the Aadhaar-based PDS, the politics of the artefact firmly locate the risk of diversion at the last mile, where ration dealers are prevented by Aadhaar-based transactions from siphoning off PDS goods.

Research has, however, questioned this assumption. Already during my fieldwork in Kerala in 2011–2012, several recipients pointed out how substantial diversion occurs *before* goods arrive at the ration shop, in the stages of storage and transportation that characterise the PDS supply chain. Concerns with a 'border mafia' (The Hindu, 2013), diverting goods at the border with the neighbouring state of Tamil Nadu, were vocally raised; this has implications for ABBA, because if diversion occurs *before* goods make it to ration shops, a system that monitors ration shops alone may have limited effectiveness. If so, the risk for ABBA is to concentrate monitoring away from parts of the distribution system that, such as transportation of goods through internal borders, are highly prone to diversion.

The second question is what happens to Ayanka. This may be easily sorted; Ayanka is here with Deva, who can authenticate without issues and collect the ration for both. But this is not always the case, and indeed the question moves to what happens to those who unlike Ayanka, do not find an alternative way to collect their rations. It is to them that the notion of legal injustice, *injustice resulting from the conditionality of legal rights and entitlements to digital identification and authentication*, applies.

The Anatomy of Legal Injustice

As of the 1948 Universal Declaration of Human Rights, food is a universal right. In the words of the UN Special Rapporteur on the right to food, it is defined as 'the right to have regular, permanent and unrestricted access – either directly or by means of financial purchases – to quantitatively and qualitatively adequate and sufficient food corresponding to the cultural traditions of the people to which the consumer belongs, and which ensure a physical and mental, individual and collective, fulfilling and dignified life free of fear'. According to the World Food Summit (1996), food security exists when 'all people, at all times, have physical and economic access to sufficient, safe and nutritious food that meets their dietary needs and food preferences for an active and healthy life'.

Food security programmes, such as India's PDS, are designed to ensure fulfilment of the right to food. The PDS covers both quantitative and qualitative dimensions of such a right: the National Food Security Act (NFSA), which is in force from 2013, legally entitles up to 75% of the rural population and 50% of the urban population to receive subsidised food grains under the targeted PDS, meaning about two thirds of the total population. Persons eligible under the NFSA are entitled to '5 Kgs of food grains per person per month at subsidised prices of Rs. 3/2/1 per Kg for rice/ wheat/coarse grains' (NFSA, 2013). High coverage is a distinctive marker of the NFSA, which in addition to the PDS establishes entitlements to nutritious food for children, pregnant and lactating women, and maternity entitlements in cash; the marker of high coverage is functional to expanding the right to food to a substantial part of the Indian population (Khera, 2011b; Sen & Himanshu, 2011).

On the qualitative side, state-level reforms have pursued nutritional outcomes. A problem reported by Reetika Khera, professor of Economics at the Indian Institute of Technology (IIT) Delhi, consisted in the food basket composition induced by the PDS. In a study of the state of Rajasthan she found that the programme, centred on the provision of rice, had the effect of shifting user consumption away from more nutritious coarse cereals (Khera, 2011a). In response to the issue, different states have provided additional food items through the PDS; these include millets and other coarse grains, pulses and lentils, as well as edible oil and biofortified foods (Hindustan Times, 2022). Overall, such measures aim at addressing malnutrition beyond hunger, and research indeed demonstrates their positive impact within states that actively reformed the PDS (Khera, 2011b; Drèze & Khera, 2017).

India's approach to welfare directly relates the PDS to the right to food. The programme is designed, implemented and reformed with specific attention to such a right. But the question is, how is an individual's right to food affected when biometric authentication is introduced?

Let us first respond on the basis of ABBA. With biometric verification, the person uses the ration shop's biometric reader to have their fingerprint recognised. If the system functions correctly, entitled people authenticate (and non-entitled people do not) – and are therefore able to access their rations. This is predicated on them having a correctly registered record in the Aadhaar database. But if such a record is missing, incorrect or authentication does not function, rations cannot be provided, which impairs the person's ability to access food. It is this conditionality, making the right to food conditional to digital identification and authentication that constitutes the heart of legal injustice.

As per its definition, legal injustice can occur at two levels. The first one is *identification*, concerned with how information on the person is established. Correct identification in ABBA is predicated on registration of user details in Aadhaar's Central Identities Data Repository (CIDR), *and* correct linking of such details to the state's ration card database. Incorrect identification has to do with issues in the same process: wrong name spellings, faulty registration of biometric details or, crucially, absent or wrongful seeding with ration card details. Lying at the deeper layer of digital identity, beyond the ration shop interface that users routinely encounter, identification faults can be hard to detect. In Karnataka, some users became aware of compulsory Aadhaar linkage only when trying, and failing, to collect their rations; during our 2018 fieldwork, a *seva kendra* owner recalled people sharing concern of what would happen of their rations if the system did not work.

But a second, more visible level is that of *authentication*. It is here that Ayanka's access journey stops, and that she needs Deva to come in. The reason Deva can step in for her is that with the ration card being a household-based document, other members registered in the same household can collect the ration. However, such a solution is not available to everyone. The consequences of denial of food rations can be extreme; in the state of Jharkhand, where a high number of ration cards were cancelled due to not being linked to Aadhaar, reports were made on hunger deaths associated to lack of access to rations (Drèze et al., 2017). Some of these ration cards, Singh (2019) notes, have been cancelled for unclear reasons, which not even the bureaus in charge of the PDS have been able to explain.

Denial of rations to entitled users is the epitome of legal injustice. For many beneficiaries, access to the PDS is the main route to fulfilling the right to food. System failures impair their ability to avail this right, whose universality is replaced by conditionality to correct identification and authentication. At its core, the anatomy of legal injustice lies in the conditionality that it introduces.

The Numbers of Legal Injustice

Faced with legal injustice, we may want a critical memorandum of how digital identity systems are linked to development objectives. In the orthodoxy of digital ID for development, identity is an enabler, and not a disabler, of rights; digital ID facilitates access to services through which essential rights, from life to food and shelter, are pursued. As then Vice President for Global Practice Solutions at the World Bank stated, in 2013, after a lecture by Aadhaar's founder Nandan Nilekani, 'neither dignity nor rights are possible without an identity' (Centre for Human Rights and Global Justice, 2022: 8).

Nevertheless, legal injustice does occur, and it extends well beyond schemes as the PDS. Published in 2022, the report by the Centre for Human Rights and Global Justice titled 'Paving a Digital Road to Hell: A Primer on the Role of the World Bank and Global Networks in Promoting Digital ID' illuminates the functionality of digital ID systems in excluding entitled users from essential services, also pointing at a lack of documented evidence of such systems' benefits (Centre for Human Rights and Global Justice, 2022: 13). A counterargument comes, however, from proponents of ID for development. Exclusion, it is argued, affects a limited number of cases, which can – with time and costs depending on the case and system – be fixed to guarantee everyone's rights. As further noted by Drèze et al. (2017: 58), starvation deaths in Jharkhand prompted a central government order for Indian state governments to ensure that all users failed by ABBA could buy PDS rations through an 'exemption register'.

The question becomes, as a result, 'how large' the exclusion issue is, and whether it is of a manageable size across countries and organisations using digital ID. It should be noted that exclusions from the PDS have been a problem since the shift to a targeted system, which in 1997 left a high share of needful households in urgent need to prove their poverty status. Difficulties in proving such a status, combined with a capped share of BPL subsidies assigned to each state, led to what Swaminathan (2002) referred to as 'excluding the needy'; a system where, unlike the previous universal PDS, substantial shares of needy users were left out by design. The good performance of Tamil Nadu, the only state where a universal PDS persisted, has been brought up to corroborate the argument that narrow targeting has been detrimental to the PDS and its users (Swaminathan, 2002, 2008; Khera, 2011b).

With its enforcement of biometric authentication, ABBA crystallised the authorisation-authentication nexus associated to such exclusions. In spite of the numerous quantitative studies of digital identity systems and their richness in documenting exclusions, the *size* of exclusions from a given programme is relatively difficult to demonstrate. Scholars have run exclusion estimates in India's Aadhaar-based PDS; in a study of Jharkhand, Muralidharan et al. (2020) show a 10% reduction in benefits for the 23% users who had not linked an Aadhaar card to benefit rolls, of whom 2.8% received no benefits at all. Such findings echo Drèze et al.'s (2017) study of ABBA in 32 villages of Jharkhand; about 7% of the surveyed households did not have a 'POS-able member', which the authors defined as a member able to use the ration shop point-of-sale machine in online mode.

Muralidharan et al. (2020) partially support the thesis of 'pain without gain' by Drèze et al. (2017). Such a thesis argues that ABBA in Jharkhand has resulted in pain for ration dealers and users, without the system resulting in the expected gain of reduced corruption. The point has, however, been debated between the two author teams. Muralidharan et al.'s (2020) argument was, on the one hand, centred on 'balancing corruption and exclusion', sustaining that ABBA – while indeed resulting in the exclusions produced by conditionality of PDS access to Aadhaar – was at the same time effective, along with policy reform, to combat diversion of goods from the PDS. Such a balance is in itself the rationale for introducing biometric authentication; as a ration dealer powerfully suggested, during our Karnataka fieldwork, exclusions are 'the cost to bear' for a better system for all.

On the other hand, the ability of ABBA to fight corruption has been seriously questioned. A rejoinder on the idea of balancing corruption and exclusion was published by Drèze et al. (2020a) on the blog 'Ideas for India'; what Drèze et al. (2020a, 2020b) dispute is the view that ABBA is of use in reducing diversion from the PDS. The argument has two components. First, the type of fraud that ABBA combats is not the *quantity* fraud (meaning the detraction, partial or total, of ration quantities from users) that prevails in the PDS. ABBA combats an *identity* fraud, committed by non-entitled users who falsify their credentials, for example with bogus ration cards. But in ABBA, as powerfully illustrated by Hartej Singh Hundal, A. P. Janani and Bidisha Chaudhuri, ration disbursal is *distinct* from authentication, meaning that successful authentication does not imply disbursal of the right quantity of goods (Hundal et al., 2020). With ABBA on the other hand, Hundal et al. (2020) report, users can be successfully authenticated without necessarily being disbursed their correct quota by the ration dealer. If so, ABBA leaves untouched the main type of fraud that it is designed to address.

Secondly, and echoing what stated before, ABBA monitors specifically the transactions that take place *in the ration shop*. Its focus on the last mile reflects a philosophy that makes the ration dealer the object of monitoring; as such, it is not planned to target earlier stages of the supply chain, such as the storage and transportation where opportunities for corruption remain. This deprives the system of an essential capability to monitor stages where diversion is reported to be highly present, as noted in the 'border mafia' concerns raised in Kerala during my research (Masiero, 2014: 159–160). At the same time, argue Drèze et al. (2017), ABBA potentially creates new incentives to ration dealer corruption by increasing their transaction costs, associated with the operations to lead every month just to run their ration shops. This point shifts the debate away from the core question on 'how many households' are excluded from ABBA, and questions whether a 'balance', with corruption check on the one side and exclusion of users on the other, exists at all.

My qualitative work does not explore the econometric side of the debate. But there is an issue that the notion of legal injustice, here developed and substantiated, offers the conceptual tools to address. Exclusion of entitled users is a *product* of legal injustice; with essential rights being not universal anymore, but subordinated to correct digital authentication and identification, it is materially possible for people to be excluded from the PDS. Rather than an incident, such exclusions are intrinsic to the system's substance, which makes entitlements conditional to correct identification and authentication of its recipients. To better understand this substance, we need to venture into the socio-technical features of what we define as the *politics of anti-poverty artefacts*.

The Politics of Anti-poverty Artefacts

Machines, argued political philosopher Langdon Winner (1980), embody specific forms of power and authority. Few examples are more renowned than that provided in Winner's seminal text, 'Do Artifacts Have Politics?' in 1980. Referring to 'technical arrangements as forms of order', Winner points to the low height of overpasses on the parkways to New York's Long Island, noting how we are prompted to interpret details of form as devoid of political meaning. Long Island's overpasses are, however, deeply imbued with such political meaning, Winner argues. He illustrates how master builder Robert Moses, responsible for the design of bridges, devised them with the purpose of impeding the passage of buses, most frequently used by poor people and Blacks. The classist and racist valence of Moses' bridges grew to become a core symbol of political artefacts, and indivisibly from that, of the harm suffered from targeted people as a result of such politics.

Anti-poverty artefacts, I argue here, are just as political as Moses' bridges. By *anti-poverty artefacts* I mean all artefacts that, in more or less material forms, participate in the design and implementation of anti-poverty policies (cf. Masiero & Prakash, 2015). While anti-poverty policy is often formulated at the national level, specific provisions can exist, such as the Green Card Scheme formerly adopted in Karnataka to enhance food subsidies for the poorest of the poor (Mooij, 1998). National policies are, at the same time, influenced by supranational directives and by the global development agenda, especially that articulated in the Sustainable Development Goals (SDGs) and the targets associated to them. All artefacts participating in anti-poverty policies – whatever their nature, and the level at which they operate – qualify as the *anti-poverty artefacts* that influence, in more or less direct ways, the lives of recipients.

As we use this definition, we have already encountered a few anti-poverty artefacts in this book. The ration card that affords people's access to the PDS – and whose absence denies it, as it happened to Aisha – was the first one we met. Let us keep in mind that artefacts do not need a digital, or at large technological, component to qualify as such; a card, which embodies people's poverty status and therefore their entitlements, plays a major role in anti-poverty policy. Think again of Aisha's story, and of her frustration in being unable to access the PDS as she lacks her ration card; not only does the card participate in the programme's implementation, but it is a material, essential requisite for users to access the provisions that the state's food policy has designed for them.

Karnataka's use of ABBA has thrown us deeper in anti-poverty artefacts. The very process of accessing rations through Aadhaar-based authentication, in turn linked with the state-level ration card database, is populated with them. Rather than operating in isolation, anti-poverty artefacts sort their effects in combination with each other. As noted again by Drèze, ABBA requires 'multiple fragile technologies to work at the same time: the PoS machine, the biometrics, the internet connection, remote servers and often other elements such as the local mobile network' (cited in The Wire, 2016). Beyond the fragility of the system, the point is on the concerted action behind anti-poverty artefacts, which work in combination with each other in producing, or seeking to produce, their intended policy outcomes.

This brings us back to the point on substance raised with reference to exclusions from ABBA. Exclusions are an unintended outcome of the system, and one that policymakers seek to avert with tools such as the 'exemption register' prospected for Jharkhand. The question is, however, on the politics that systems of the type of ABBA, which subordinate access to biometric authentication, embody within themselves. In other words, *what is the politics of the anti-poverty artefacts* that ABBA seeks to advance?

Answering this question requires a reminder of the two main types of errors – defined, with Devereux and Sabates-Wheeler (2004), as *exclusion* and *inclusion* errors – in which targeted social protection systems can incur. Targeted social protection systems, it should be reminded, exist in virtue of the presence of a target population entitled to access. An *exclusion error* occurs when genuinely entitled users are excluded from the system. Aisha who is not given a ration card and Ayanka who cannot authenticate with ABBA are both victims of exclusion errors. Conversely, an *inclusion error* occurs when users who are not entitled to a system are erroneously included in it. Owners of bogus ration cards, presented by the Justice Wadhwa Committee Report (2010) as a major problem of the PDS across states, enable an inclusion error by affording non-entitled users to illicitly access the system.

The biometrically enabled PDS enforces a very specific policy measure. On the one hand, its action is explicitly tailored to combat the inclusion errors that have plagued the system for years. Denial of access to the non-entitled is written into the functioning of the technology; a person without a ration card, or a person which the system does not recognise as a registered user, cannot be made to access the PDS. In its virtue, the technology carries the message that access is to be restricted to validly identified and authenticated PDS users. While extensions can be added into the system, such as the exemption register prospected for Jharkhand, the heart of the technology makes access conditional to successful authentication, equating its failure with non-entitlement.

The system's architecture answers the question on the artefacts' politics. Designing a technology that bars access to the non-recognised, but takes no provisions for the erroneously excluded, the artefact prioritises the fight to inclusion errors, widely recognised in national policy as responsible for diversion and leakage (cf. Bardhan, 2011). The history of the PDS gives a strong rationale for such a policy; with the transition to a targeted system in 1997, and the spike in diversion that ensued from it, securely identifying genuine beneficiaries became prioritarian (Mooij, 1999). While such an argument has been questioned on the grounds that exclusion errors were at least equally pressing, the purpose of tailoring the system to fighting wrongful inclusion has shaped the technology into what it is today: a technology making a universal right conditional to digital identification and authentication.

Two notes of caution must be made on this argument. First, biometric recognition (and the database of biometric details at the basis of it) is *not* a fundamental premise of the fight to inclusion errors. Ration cards, which remained starkly non-digital until quite recently, play the same role, subordinating people's access to the system to recognition of the person as a valid beneficiary. The fact that in the non-digital PDS recognition was based on the person's name and photo, rather than their biometric details, does not change the purpose of the technology. Biometrics only inscribe the same politics of physical ration cards into a digital artefact, made to enhance accuracy of recognition.

Second, and crucial, the politics of anti-poverty artefacts in the PDS does not need ABBA. Shortly after Aadhaar's launch as Unique Identity Project (UID), Khera (2011c) already noted how smart cards or food coupons also provide forms of 'portable' ID, and that these methods can be significantly more effective in tackling the exclusion errors that UID does not combat. She later examined (Khera, 2018) diverse ways through which biometric authentication can be performed without ABBA: similarly, Allu et al. (2019) reviewed different alternatives for biometric authentication in the PDS. They classify methods on 'mode of authentication, biometric or non-biometric, source of authentication, central database connected through internet or locally stored data in the point of sale device (ePoS), frequency of authentication, at every transaction or once for a predefined number of transactions' (Allu et al., 2019: 30). Their study of eight states shows how most benefits in performance are associated with the use of biometrics during *identification*, while biometric authentication at ration shops, as it is in ABBA, is not associated with more effective systems. Such research illuminates how ABBA is only one way to enact food security policy, and not the most effective one that research reveals.

A special note should be made for the state of Tamil Nadu, where the continuation of a universal PDS is combined with an artefact, a biometric smart card with a QR code that presents yet another alternative to ABBA. In their comparison of the PDS in Karnataka and Tamil Nadu, Hundal et al. (2020) note that in Tamil Nadu, smart cards maintain a digital trail of all transactions, using a smart card reader in lieu of the Aadhaar-based point-of-sale (Khera, 2018). Biometrics here are not recognised at authentication, but only at the moment of Aadhaar seeding of ration cards. Hailed for its benefits, for instance delinking ration collection from the person's fingerprints (important, for instance, for elderly recipients with mobility issues), the Tamil Nadu system has also been researched in the light of several issues. As deepened in Chapter 8, Carswell and De Neve (2022) found smart cards to be associated to information gaps, leaving users unable to prove that monthly ration collection has taken place. The politics of the artefact here escapes fingerprinting, being reflected in the new design that the smart card presents.

Even more fundamentally, the politics of anti-poverty artefacts can be performed with systems that do not involve Aadhaar, but different systems of biometric recognition. In a field study of Karnataka in 2014–2015, Amit Prakash and I researched a pre-Aadhaar system that collected biometric details for PDS recipients, matching them with the details collected in the state's ration card database, known as Ahara. Selected ration shops in the state had been provided a point-of-sale machine which was augmented with a weighing scale, where ration dealers weighed goods associated to each ration cardholder. While antecedent to ABBA, the system played its same function, and in August 2014 it was described to us by the former Karnataka's Food, Civil Supplies and Consumer Affairs Secretary as a major driver of long-term change in the state's fight to corruption (Masiero & Prakash, 2015, 2019; Prakash & Masiero, 2015). As deepened in Chapter 6, this system made no use of Aadhaar; its goal, as the Secretary stated it, was that of reinforcing the PDS, rather than supplanting it with cash transfers or other measures.

It should be noted that, in the legal history of Aadhaar, the constitutionality – or not – of the scheme has been largely debated. In 2012, Justice K.S. Puttaswamy – a retired judge of the High Court of Karnataka – filed a writ petition that challenged the constitutional validity of the Aadhaar project, primarily on the grounds of infringement of the right to privacy under the Constitution of India (Global Freedom of Expression, n.d.). In response to the Government's proposal to make Aadhaar mandatory for access to Government services and benefits, other petitions were also filed, challenging different aspects of the scheme. In an interim order dated 11 August 2015, the Supreme Court noted that collection of demographic and biometric data by the Government was questioned primarily on the grounds that it breached Indian's constitutional right to privacy. But the Attorney General, on behalf of the Union of India, argued that the Indian constitution does not grant specific protection of the right to privacy (Global Freedom of Expression, n.d.). These arguments reflected an unresolved contradiction in the law around the status of privacy in India. Accordingly, they referred the discrete question of whether the right to privacy was a constitutional right under the Indian Constitution for adjudication by a larger constitutional bench of judges. On 24 August 2017, a larger nine-judge bench of the Supreme Court decided that the right to privacy is protected under articles 14, 19 and 21 of the Constitution. The Court also laid down a test to be applied in adjudications of whether invasions of this right were justifiable or proportionate to other state interests.

The decision had important consequences for Aadhaar, as the matter returned for adjudication to the five-judge bench that had been hearing it. On 26 September 2018, the Supreme Court upheld the validity of Aadhaar, and its compulsory use in certain State welfare schemes as on the grounds of its functionality for the delivery of welfare benefits. But after applying the test for privacy, the judgement struck down Section 57 of the Targeted Delivery of Financial and Other Subsidies, Benefits and Services Act ('Aadhaar Act') which enabled private entities to use Aadhaar for the delivery of services. In addition, the Court directed that schools, banks and telecommunication companies cannot make Aadhaar mandatory for accessing their services.

It is the politics of the artefact that informs my point on the substantial, rather than accidental, nature of legal injustice. While exclusions of entitled users are largely framed as a side effect, *they are directly produced by how the technology is designed*. A technology that writes an anti-inclusion error policy into the artefact – rather than a policy that averts exclusion errors – produces the outcomes that Aisha, Ayanka and many others suffer, and that only augmentation of the original artefacts with extra tools can address. Legal injustice, which makes universal rights conditional to correct identification and authentication, is the epistemological underpinning of such artefacts.

The Biometric Artefact in Action

My conception of legal injustice emerged in the context of India's PDS. Yet the embeddedness of the authorisation-authentication nexus into biometric artefacts, in and beyond anti-poverty programmes, invites a broader reflection on the matter. If the substance of the artefact is that of combating inclusion errors at the expense of exclusion ones, this may affect authentication systems well beyond food security and its related technologies. In the section below I focus on the artefact politics reflected in two cases: the biometric database of refugees in Kenya, where double registration resulted in denial of citizenship rights for those affected and the digital ID system in Uganda, which has been implicated in large-scale exclusions from healthcare and cash entitlements.

Kenya: From Humanitarian Assistance to Double Registration

The work of Keren Weitzberg, currently a Senior Lecturer and Fellow at the Institute of Humanities and Social Sciences of Queen Mary's University, London, exerted a deep influence on the understanding of biometric artefacts presented in this book.

A historian whose research focuses on biometrics, mobility and national inclusion, Weitzberg is the author of 'We Do Not Have Borders: Greater Somalia and the Predicaments of Belonging in Kenya' (2017), a book that examines the historical factors of the perceived foreignness – and associated abuse – long suffered by persons of a Somali ethnicity in Kenya. Her rigorous archival analysis shows the long presence of Somalis in present-day Kenyan borders, as well as the ways in which identity-making resulted in systemic violence on them. Throughout colonial history, Kenyan borders were arbitrarily designed; it is, as such, consequential that communities close to the border share cultural and ethnical backgrounds, resulting into a situation of significant border porousness (Weitzberg, 2017; Haki na Sheria, 2021: 11).

Weitzberg's (2017) book offers a clear picture of such porous borders. One of the main reflections of this is the movement of people across the border, which over time produced important consequences on the delivery and shaping of humanitarian assistance in Kenya's refugee camps. In the early 1990s, the outbreak of the Somali civil war resulted in massive fleeing from the country, and coincided with one of the numerous periodic droughts in north-eastern Kenya, which led many Kenyans to flee to refugee camps inside the country (Weitzberg, 2020b; Haki na Sheria, 2021). With a large influx of people in the camps, the UNHCR staff in charge of humanitarian assistance struggled to distinguish people who qualified as refugees as they fled Somalia, from national Kenyans who, while in desperate need of vital supplies of food aid, education and medical service, did not qualify as refugees. Many Kenyans turned up at camps, claiming Somali citizenship to access the food and primary aid they needed for survival (Weitzberg, 2020b; Haki na Sheria, 2021; The Engine Room, 2023).

The report 'Biometric purgatory: how the double registration of vulnerable Kenyan citizens in the UNHCR database as left them at risk of statelessness', authored by the Kenyan non-governmental organisation Haki na Sheria, is one of the most comprehensive sources on *double registration*, a term used to indicate how people, registered in the UNHCR database of refugees, became unable to apply for a Kenyan national ID card. Haki na Sheria, whose story of work with double registered people is detailed in Chapter 7, was founded in Garissa, a north-eastern Kenyan county bordering Somalia and characterised by high securitisation, with people being unable to leave or enter the county without a valid proof of ID (Haki na Sheria, 2021: 4). When applying for a national ID, residents of Garissa are among the most targeted by *vetting*, a process that refers to the screening of applicants for a national ID card to verify their details. Residents of border regions, notes again Haki na Sheria (2021: 11), are subjected to vetting as a result of the suspicion that accompanies their status; applicants need to prove that they are really from Kenya, and not from other countries. As Haki na Sheria (2021, 2023) describes, the practice of vetting can be oppressive and humiliating, and it turned the production of an ID card from a universal right to a right conditional to a vetting committee's approval.

The biometric artefact, consisting in this case of the UNHCR database that included the biometric data of people registered as refugees, is supposedly built with the purpose of protecting vulnerable people. It can be said that humanitarian biometric artefacts mirror, by all means, the anti-poverty artefact logic; through people's biometric credentials, the database marks the refugee status to which food, shelter and aid are made conditional. The UNHCR first introduced digital biometric registration in Dadaab, Kenya's largest refugee camp, in the mid-2000s, and later shared the refugee database with the Kenyan government in 2012 (Haki na Sheria, 2021: 4). The availability of such data makes the fingerprints of people registered as refugees by the UNHCR readable by Kenya's authorities, notifying, at the time of applying for an ID card, if the person already holds refugee status as a result of prior registration.

But it is, Kenya's history tells, exactly this logic of data-for-humanitarianism that the UNHCR biometric database defied. Kenyans who claimed refugee status, to access goods as vital as food and shelter, became subject to double registration; registered as refugees in the UNHCR and Kenyan government databases, they are, as a result, turned down at the Kenya National Registration Bureau when applying for a national ID card. The rationale of the norm lies in the body of the artefact: interoperable with the National Bureau of Registration, Kenya's Automated Fingerprint Identification System (AFIS) includes refugee biometrics in such a way that, when the person applies for an ID card, the refugee status associated to the fingerprint results in a negative outcome of the application. Weitzberg (2020b) shows how this affects double registered individuals, for instance M.H., a man in his 20s who, unable to access education or formal employment due to lack of an ID card, has opened a street kiosk:

> "If I want to open an M-Pesa line," he said, referring to the popular mobile payment service, "I can't register without an ID. If I get a bit of money, I don't have a place to put it. It's necessary for me to keep it at home. And at home, it can be lost. I don't have any kind of [bank] account. (Quoted in Weitzberg, 2020b: n.p.)

One of the most striking aspects of double registration is that people, severely affected by inability to obtain a national ID card, were often unaware of the consequences of refugee registration. Haki na Sheria (2021) deepens this point – in fact, many found out about their predicament only when turning to the National Bureau of Registration, and seeing their application for an ID card turned down. For many, registered during infancy by their family members or carers, this only happened when turning 18 years old, the legal age for obtaining a national ID card (Haki na Sheria, 2021: 27). Showing up at the National Bureau of Registration for an ID card application, many received the news that their fingerprints were in the refugee database, barring them from having an ID and effectively putting them at risk of statelessness (Haki na Sheria, 2021; Mutung'u, 2021).

Haki na Sheria's (2021) report is central in voicing double registered people, which amount to around 40,000 over time (Haki na Sheria, 2021: 4). Many are under the age of 40, and had their fingerprints registered during childhood; many do not, in fact, remember the act in which their fingerprints were taken (Haki na Sheria, 2021: 28). 'My mother died during my birth', tells one double registered person interviewed in the Haki na Sheria report: 'my aunty, who brought me up with her children, tells me that we went to the camps with her two sons during the drought' (Haki na Sheria, 2021). Others share how they went to the camps for food aid, leaving just after obtaining it; others again note how they registered to increase the head count of a family member, an increase which would benefit the family as a whole. But the fingerprints of the same people, notes again Haki na Sheria (2021: 28), never left the database, and became cross-checked with the national register of persons in a way that resulted into structural denial of an ID for them. The binary nature of the biometric artefact, notes The Engine Room (2023: 72), makes it even more difficult for the wronged parts to contest the decision made on them.

The work of Grace Mutung'u, an advocate of the High Court of Kenya and research fellow at the Centre for Intellectual Property and Information Technology at Strathmore University, further illuminates many consequences of double registration for affected people. Mutung'u (2021) focuses on the legal aspects of the National Integrated Identity Management System (NIIMS), launched by the Government of Kenya in March 2019: known as *Huduma Namba*, meaning 'service number' in Swahili, the programme tried to centralise identity schemes, issuing a unique number to citizens and registered foreigners in Kenya. The programme's structure echoed India's Aadhaar under multiple aspects (Gill, 2020) and similar exclusionary issues, note Mutung'u and Rutenberg (2020: 348), came along with its proposal, as people without a primary identity document could not register with it. This essentially excluded double registered people from the Huduma Namba fabric; found to be in violation of the country's Data Protection Act, the scheme was declared illegal by Kenya's High Court in 2021 (Open Society Justice Initiative, 2021). At the time of writing, a new initiative termed Unique Personal Identifier (UPI) has been piloted by the Kenyan government; associated to an identifier named *Maisha Namba* (life number), the UPI aims to build a unique and verifiable digital identifier, with features that have been argued to closely resemble Huduma Namba (Abuya, 2023; Access Now and eight other signatories, 2023). Discussed in Chapter 7, the case of Maisha Namba again interrogates the plight of those who suffer from denial of an ID card in the first place, and questions around it pertain to issues of exclusion and discrimination that the story of Huduma Namba, just few years before, had opened (Access Now and eight other signatories, 2023).

The double registration of people in Kenya illustrates how legal injustice, rather than an unintended consequence of the biometric artefact, can be directly inscribed in it. The right to a national ID is subordinated to a form of identification that the refugee database, as it is designed, denies; the person cannot be recognised as a Kenyan citizen, due to the refugee status that the database associates to them. In response to this, civic organisations have exerted pressure for the government to begin a process of deregistration of people as refugees. While ID cards started to be released to double registered people in January 2022 (Haki na Sheria, 2022), the artefact is inspired by the subordination of citizenship to the lack of a previously registered refugee status, exerting the conditionality that produces legal injustice. Chapter 7 will expand on the activism of Haki na Sheria, and its pivotal role in reversing legal injustice for double registered people in and beyond Garissa county.

The injustice implicit in double registration may arguably not surprise the reader who comes to this book with knowledge of inner dynamics of digital humanitarianism. An authorisation-authentication logic is firmly inscribed in biometric artefacts used in the humanitarian context, a point that, reflected in more recent surveillance studies literature, holds largely beyond the Kenyan case. Paradigmatically, in 2019 the World Food Programme (WFP) introduced its partnership with the private tech software company Palantir, with the goal to 'deliver life-saving assistance using data' (WFP, 2019a). In launching the partnership, WFP executives declared:

> The sheer scale of WFP's operations, assisting some 90 million people in about 80 countries, means that even small efficiencies in operational and supply chain management can lead to dramatic savings. (…) Making this data accessible across the organization will help WFP become even more efficient in multiple programme areas, including cash-based transfers, supply chain optimization, and nutritional requirements. (WFP, 2019a, n.p.)

As Martin (2023) notes in discussing the WFP-Palantir case, there is grounds to argue that surveillance firms may be exploiting humanitarian crises to *aidwash* their technologies and the services connected to them. If this is the case, the data-feeding functionality of biometric artefacts is subservient to the logic of aidwashing, where the artefact politics crystallised in biometrics reinforces the purposefully built humanitarian image of high tech providers. As the point is deeply connected to management of user information in humanitarian programmes, we will return to this point in Chapter 6.

Uganda: Digital ID and Exclusion From Entitlements

In Kenya, it is a humanitarian biometric artefact that results in denial of a national ID card to thousands of residents, depriving them of access to essential services. With India's PDS we have conversely seen the work of biometric artefacts at the state level, with the diverse consequences predicated on making essential services conditional to digital ID. Such consequences, as noted from countries that have more recently introduced foundational ID systems, can be transformative for national services. In June 2021, the Centre for Human Rights and Global Justice of New York University, the Initiative for Social & Economic Rights, and Unwanted Witness published 'Chased Away and Left to Die: How a National Security Approach to Uganda's National Digital ID Has Led to Wholesale Exclusion of Women and Older Persons', a report centred on the transition of Uganda's ID system to an architecture that makes essential service provisions conditional to having a digital ID.

Uganda's digital ID system, notes the report, long relied on functional ID systems including 'voter's ID cards, drivers' licences, baptism cards, graduated tax slips, and letters from the LC1 Chairperson' (2021: 24).[1] When the national identity system, popularly referred to as *Ndaga Muntu* ('identity card' in the Luganda language), was introduced in 2014, its introduction was accompanied by the National Identification and Registration Authority (NIRA), a dedicated agency established to manage the novel ID database (McDonald, 2022). The mandate of NIRA was 'to collect data, including biometric fingerprint scans and facial photographs; and to issue a unique national ID number and a national ID card to those who successfully pass a citizenship verification process' (Cioffi, 2023). Associated to promises of expedition of service delivery and inclusion through digital ID, the Ugandan system involved three connected artefacts: a central database called National Identity Register (NIR); a National Identity Card (NIC); and a National Identity Number (NIN) associated to each person with their credentials (Centre for Human Rights and Global Justice, 2021: 25).

The report in point was, however, one of the first sources to comprehensively problematise the ability of the Ugandan ID system to deliver on its own promises. Wholesale exclusions, notes the report, are in themselves problematic; based on the Government's own data and other official sources, the Centre calculated that 'anywhere between 23% and 33%' of Uganda's adult population had not received a NIC at the time of the study (2021: 9), an issue combined with the cumbersome and

---

[1]The LC1 Chairperson is the chairperson of the local council, who often is the primary point of contact for local community members (Centre for Human Rights and Global Justice, 2021: 27).

costly process of replacing lost or damaged cards (2021: 10). National law supports the need to prove identity through NIC to access basic services; the Registration of Persons Act, promulgated in 2015, serves as a legal basis of the lawfulness of demanding production of a NIC or NIN for accessing services as essential as healthcare and social security. Enforced at multiple access points for service provision, the act makes access to basic entitlements conditional to digital identification and authentication, against the backdrop of wholesale exclusion from card ownership that the report brought to light.

The same report relies on qualitative fieldwork, conducted with victims of exclusion related to NIC, to narrate stories of exclusion from two essential systems: healthcare for women, especially pregnant women, and recipients of social security for older persons, particularly those living in poverty. On the latter, Uganda presents a universal old age pension – referred to as the Senior Citizens' Grant (SCG) – which has been hailed as a highly successful route to social service provision. Kidd (2017) finds that the scheme, which had 125,000 beneficiaries by 2016, is associated to significant reductions in poverty among beneficiaries and in the number of households experiencing hunger. Indirect benefits have affected multiple sectors of the society; these have included improved diets, investment in productive activities by beneficiaries and increases in employment among working age members in pensioner households. In addition, the affected pensioners have reported to experience greater dignity, leading to re-integration into their communities (Kidd, 2016: 3); a Value for Money audit report by the Auditor General in 2021/2022 confirmed the strong impact of the programme on recipients (Office of the Auditor General, 2022). Under all such aspects, the programme seems an example of *transformative* social protection (Devereux & Sabates-Wheeler, 2004), whose effects transcend the beneficiary to affect their economic, community and societal ecosystems.

And yet, it is exactly on beneficiaries of healthcare and the SCG that the report by the Centre for Human Rights and Global Justice (2021) has documented rampant exclusions. Lack of a national ID card, which alone affects a substantial share of the population, is only one of the routes to exclusion documented in the report, which features stories of pregnant women being denied access to healthcare clinics due to inability to produce a NIC (2021: 10). Entitlement to the SCG is conditional to accurate reporting of age from beneficiaries; this is especially problematic in Uganda, where erroneous reporting on age as recorded in the NIR affects an estimated 40,000 persons (2021: 42). An additional requirement relates to the correctness of name and age in the lists produced by NIRA and validated by the Programme Management Unit of the Ministry of Gender, Labour and Social Development (ibid.). With the periodic production of such lists, wrongful details can impact people's ability to receive cash for a long time, leaving them factually unable to take direct action for correcting mistakes.

What is crucial in the Ugandan case is the direct implication of the biometric artefact in the exclusions documented by the report in point. It is again the SDG to illustrate the role of biometric credentials in the matter. Older persons, it is noted, need to be physically present at the point of collection of their subsidies, to ensure correspondence of their biometric credentials with their records (and verify the fact that they are still alive); while the Expansion of Social Protection programme of the Ministry of Gender, Labour and Social Development in principle allows for alternate recipients, in practice this is very difficult, as stories collected in the report reveal (2021: 42–44). As noted for biometric collection of food subsidies from India's PDS, the requirement of in-person collection is especially burdensome for immobile persons, making the matter particularly difficult for the elderly – exactly the category targeted by the SCG. The unity of biometric credentials with the act of collection reflects an artefact politics that reminds again of the story of India's PDS; exclusion errors, which affect a large share of the population, are deprioritised with respect to the risk of erroneous inclusion of people in the SCG.

Such deprioritisation has been reflected in the legal matters that, over the last two years, have affected the digital ID system in the country. In April 2022, three civil society organisations – the Initiative for Social and Economic Rights, Unwanted Witness and the Health Equity and Policy Initiative – have filed a suit against the Ugandan government, stating that the mandatory use of NIC for accessing key services amounted to violation of citizens' rights (Reuters, 2022). The lawsuit especially highlighted the high vulnerability of those excluded in terms of old age, poverty, and person with disabilities, also noting the burden suffered by people with errors on NIC, whose access to key services is also limited. The organisations requested the Uganda High Court to declare that sole reliance on the national ID to access core social services is exclusionary, discriminatory and violates human rights, and to provide immediate remedies for those who have been excluded (Cioffi, 2023; CIPESA, 2023).

The verdict on the case is pending at the time of writing. An important development has, however, occurred in March 2023, when the Court accepted an amicus curiae brief from Access Now, Article 19, and the Collaboration on International ICT Policy for East and Southern Africa (CIPESA). Another amicus curiae brief on economic and social rights was submitted by Philip Alston, former UN Special Rapporteur on extreme poverty and human rights (Centre for Human Rights and Global Justice, 2023). The briefs pertained to potential human rights impacts of NIC on 'the right to privacy, the right to freedom of expression, as well as intersecting economic, social, and cultural rights, by providing experiences from national, sub-regional, regional and international levels' (CIPESA, 2023). So far, notes Cioffi (2023), exclusion has not yet been grounds for invalidating a digital ID system. In this respect, Uganda's case – beyond illustrating the impact of biometric artefacts on large-scale exclusions from key services – can offer an important legal precedent on the role of digital identity in essential service provision.

Summary

This chapter has marked the beginning of our journey into the injustices of digital ID. Starting from India's ration shops, we have reached Kenya and Uganda in examining digital identity systems that, while promising to improve people's welfare, result in denial of rights as essential as food, healthcare and citizenship. An analytical tool derived from Winner (1980), the politics of anti-poverty artefacts has arisen as a route to illuminate such injustices, studying digital identity systems in terms of the historically, geographically and socially shaped politics designed into them. It is with the same lens, and with its attention to the user, that we proceed in our journey through informational injustice.

5

INFORMATIONAL INJUSTICE

Exclusion errors directly hamper people's ability to access essential services and programmes. With its impact on human lives, *legal injustice* – as illustrated in Chapter 4 – is epitomic of the harm that failures in digital authentication and identification can generate on users. At the same time, other forms of injustice are less directly visible, due to being produced within layers of service delivery, social protection or assistance that lie behind the 'last mile' where people encounter providers. Our journey through these layers begins on a ration distribution day in south-eastern Karnataka.

'If I Can Get Rations, It's Ok'

Ankita, a middle-aged PDS user, meets us by her house near the local ration shop. The shop is open for the first ten days of the month, and in case of any variation in times – for instance, if commodity stocks on a given month are delayed – a sign is usually put up on the shop's door. On the shop's wall, a blackboard reports the monthly allocations of different food grains for users classified as Antyodaya Anna Yojana (AAY) or below-poverty-line (BPL). Ankita tells us about ration collection, which she has availed from the same shop since long before Aadhaar-Based Biometric Authentication (ABBA) was introduced. When asked about Aadhaar enrolment, she shows no exact recollection of the timing and procedure involved, or indeed of being physically present to it: 'my husband did that for me', she says.

The puzzling point of such a recollection is that Aadhaar enrolment involves registration of biometric credentials, hence requiring the individual to be present on site. Ankita does not detail such a process; she is instead keen to tell us about ration delivery, and particularly about how it happens through the biometric system that ABBA introduced. She gives a positive description of ABBA's functioning; as the ration shop's reader recognises her fingerprint, she experiences monthly ration collection as a smooth, largely problem-free experience. Sometimes, she says, the ration dealer does not give her change back when she hands in a banknote, but little can be done in response to that, she feels.

On the one hand, Ankita is extremely articulate on ration collection, which comes across as the part of the PDS that matters most directly to her. Ration collection is what she needs to provide food on her household's table, and having not experienced issues of fingerprint recognition, she is more than satisfied with how ABBA has made the process smoother than it was. What she has limited recollection of is, on the other hand, the process through which she enrolled in Aadhaar, enabling the biometric collection of rations on which her family's livelihood depends. UIDAI's legal framework is described in public sources, as well as the data protection policies applied to Aadhaar enrolees. Ankita's story, at the same time, remains blurred on the very dynamics of her registration.

Later in my research, I realised that my original question on Aadhaar enrolment was misplaced. What matters for Ankita, whose household depends crucially on subsidised food distribution, is that food rations are available every month, in a timely manner that the system, as she experiences it, mostly guarantees. Her account places less emphasis on the use that providers make of her personal data and registered biometric credentials. To my question on what happens with the data she gave out at enrolment she answers, poignantly and briefly: 'if I can get rations, it's ok'.

Ankita's answer illuminates a common position across PDS users. Chapter 4 has detailed how the introduction of ABBA has made access conditional to well-functioning biometric authentication, in turn based on the correct Aadhaar registration of users and seeding of ration cards. All these aspects are, however, a back-end to which the act that matters to users – timely and correct collection of rations – is subordinated. It is ok, Ankita intends, to go through enrolment, part with uniquely identifying biometrics and share data with providers; all of this is needed to access vital food supplies, which de-prioritises the question on how such data are treated and dealt with.

As we continue conversations with PDS users, we realise that questions on data treatment – what happens to people's data, and how are they handled under ABBA? – are often met with a blank look, and we are sometimes requested to rephrase them. A common feature of our respondents is however that of not being in the position to question UIDAI, the central government, or PDS providers on data treatment. The choice is between Aadhaar enrolment, followed by seeding of ration cards, and no reception of goods at all; people are not in the position to meaningfully question the enrolment and identification processes, as opting out of Aadhaar seeding implies interruption of access to the PDS. Being unable to question such processes makes them vulnerable to informational injustice, which in this book we define as *injustice perpetrated through the obscuration of information on use of data from digital identification*.

'Of What Use Is My Knowing?'

The centrality of *information* to discourses of socioeconomic development has been largely discussed over time. Dissections of the meaning of information, especially in the light of needs and expectations of 'development', are however far less common. In recent, groundbreaking work, Janaki Srinivasan, an Associate Professor at the International Institute of Information Technology Bangalore (IIITB), has developed important tools to read 'information' as produced within people's routinary assimilations of it. In her book 'The Political Lives of Information: Information and the Production of Development in India', Srinivasan (2022) problematises a view that depicts information as an abstract, immaterial concept, and develops the conceptual tools to read the meaning of information as entrenched in people's lived experiences of it.

Srinivasan's book starts with a quote of the former Executive Director of the United Nations Economic Commission for Africa (UNECA), noting how 'information empowers and information frees people at all levels of society, regardless of their gender, their level of education or their status, to make rational decisions and to improve the quality of their lives'. This view direly clashes, however, with the story of Palaniammal (a pseudonym), a daily wage agricultural worker whom Srinivasan interviewed in the course of her research on the Information Village Research Project (IVRP) in Puducherry. Palaniammal describes how she, as well as other agricultural workers, were sitting at home as no agricultural work was to be found near her village. When asked by Srinivasan (2022: 2) whether the village information centre would help her find other livelihood options, she noted that such centres would get her and others to learn about loans and the application process for them, but asked back: 'Of what use is my knowing?'.

Palaniammal, Srinivasan (2022: 2–3) continues, refers to the fact that people in her position do not own the assets, documents or connections needed to receive a loan, independently of their 'knowing' of how the application process works. Her position is epitomic of customers of IVRP, a village e-kiosk project based on the very idea of information for development animating ICT4D initiatives from the 1990s on. On the one hand is the deeply rooted ideology of information as a key building block of development; 'knowing', it is poignantly argued (World Bank, 1998), is essential for individuals to make the informed decisions that lead to the better quality of life alluded to by the UNECA official quoted by Srinivasan (2022: 2). But on the other hand is the reality of people who, like Palaniammal, struggle to see the use of information to build the livelihoods they need, and remain in the hopeless wait for such information to turn into useable assets.

Srinivasan's (2022) book stems from the need to conciliate these positions. To do so it relies on Bayly's (2000: 3) notion of *information order*, defined as 'a social-technological formation that is neither a completely physical infrastructure nor only the social networks in which information products circulate'. Such a concept openly questions the idea of information as a generic object, be it intended in a more particularistic (market information, information on crop fertilisers) or a more abstract way ('information' as a broader object of research). With its nature as a social-technological formation, an information order views the object of information in the light of *both* its physical infrastructure and the networks through which it is circulated; information is never, in other words, independent of the contextual settings it is inscribed in. If Palaniammal was an asset-owning, politically connected civil society member, chances are that her 'knowing' of loans would be of different use to her.

Taking Bayly's (2000) concept of information order, Srinivasan (2022: 18) modifies it to include 'the information systems, laws, procedures, documents, and records that frame such interactions, as well as the physical and social infrastructure (Elyachar 2010) that is made use of in the course of interactions'. Such a modification is inspired, among others, by Corbridge et al.'s (2005) idea of 'seeing the state' through its physically produced embodiments. In her augmented notion of an information order, Srinivasan leverages two elements of state-citizen encounters: the technologies that structure them ('information systems, laws, procedures, documents and records') and the 'physical and social infrastructure' through which encounters happen. Like Palaniammal, other users of the IVRP project conceptualise information in terms of how encounters with it occur, which in turn shape the use such information may have for them.

In reading Palaniammal's story years after meeting Ankita, several parallels come to mind. In our interview by her house, Ankita quickly glossed over information on Aadhaar registration and data use that came across as being less than important for her. The vagueness of her recollection of Aadhaar enrolment was especially striking as compared to the detail she gave on the ration collection process. Ankita spoke diffusely about ration shop opening days, the information signs put up in case of delayed ration distribution and especially the functioning of the fingerprint reader through which she collects rations. For her it is ration delivery, rather than the back-end use of biometric and demographic data, that constitutes the central element of the food distribution system.

Srinivasan's (2022) modified notion of information order is central to the point made here. In studying digital identity systems through users' perspectives, placed in physical-social realities that, such as India's ration shops, characterise people's routinary interactions with service providers, the situated character of information orders acts as a core interpretive device. Such a study starts from the very spaces where information, as for Ankita and Palaniammal, is produced or denied and it is through a situated, material vision of information that two different types of injustice, which I refer to as *opacity* and *erasing*, come to light in people's lived experiences.

Opacity: The Denial of Information

Let us go back to the story of Adeela, which we narrated in Chapter 2. Presented with the hypothetical option of receiving cash transfers instead of food rations, Adeela gave us a strong negative answer, indicating her remarked preference for the system of food rations to which she is used. People reading her story with a macroeconomic perspective may well be surprised with it. In effect, what is wrong with cash transfers? Economically they present a move that reduces the scope for leakage and, at the same time, removes the market distortion that disproportionately affects the poor (Ramaswami et al., 2014). In addition, benefits induced by cash transfers on social protection beneficiaries are well documented by research. Valid examples are found in Brazil, where the large cash transfer programme Bolsa Familia is associated to important health, nutritional and educational benefits (Neves et al., 2022), and Niger, where mobile money transfers were found to be correlated both with household diet diversity and with the number of meals had by children (Aker et al., 2016).

To understand the problem we need to dive deeper into the history that plots cash transfers as an alternative to the PDS. It is a history that begins in the early 1990s, when concerns about the exposure of the universal PDS to leakage (Ahluwalia, 1993) resulted in the transition of the country to a targeted system, where food grain allocations depended on state poverty levels. Launched in 1997, the targeted PDS deeply reshaped people's access to subsidised goods; a high share of APL users, for which only a minimum subsidy (decided on a state basis) remained, opted out of the ration shop system and resorted to the free market. As noted in Chapter 2, the transition to a targeted PDS has had extremely hard consequences for ration dealers' businesses. At the same time, it has sharply reduced fiscal expenditures, making the government's anti-poverty strategy significantly more manageable (Umali-Deininger & Deininger, 2001).

While marking a significant step in the reduction of social protection costs, the transition to a targeted PDS has been far from capable to reduce all problems. The PDS literature of the early 2000s (cf. Dutta & Ramaswami, 2001; Ramaswami & Balakrishnan, 2002) is clear on the matter. On the one hand fiscal expenditures have been reduced, making it simpler to handle one of the largest anti-poverty schemes in the world. But on the other, at least two large sets of problems emerged from the targeted PDS. First, the introduction of two categories of ration cards (which became three with the introduction of AAY in 2000) generated an incentive for people to obtain a BPL card, even when not entitled to BPL status. An independent investigation, the Justice Wadhwa Committee Report (2010), identified the incidence of fake ration cards as a major concern for the system, leading to affluent individuals obtaining BPL cards while thousands of needy families became, at the same time, excluded from the PDS (Swaminathan, 2002, 2008).

The second problem emerged, however, with ration dealers. We have noted the dire consequences that the transition to a targeted PDS had on them; with the sudden shrink of their customer basis, and the dramatic impacts such as the debt-induced suicides of ration dealers in Kerala (Suchitra, 2004), the subtraction PDS goods from ration stocks became an issue. Impoverished by a policy move that hampered their ability to run their businesses, ration dealers came to the centre of allegations of turning rationed items away from the shop and on the private market, where margins could be made. This form of corruption became, notes Khera (2011b: 1048), 'a requirement for economic survival' in the face of adversity. Systemic diversion at the ration shop level became a matter of concern to national food policy, reinforced over time by high levels of leakage from the PDS across states (Gulati & Saini, 2015).

Two caveats are to be made at this point. First, the link of both issues – poverty status assignation and ration dealers' behaviour – with the introduction of a targeted PDS is notable. Madhura Swaminathan, Professor and Head of the Economic Analysis Unit at the Indian Statistical Institute, Bangalore, has dedicated substantial research to illuminating the consequences of the targeted PDS on genuinely poor households that became excluded from the system. In her paper 'Programmes to Protect the Hungry: Lessons from India', Swaminathan (2008) draws clear links between targeting and the worsening of food insecurity in the country. Using National Sample Survey (NSS) data for India's rural areas, she notes extensive misclassification of BPL households as APL, combined with the case of states where, like in Orissa, 33% of agricultural labouring households did not own a ration card at all. The situation, Swaminathan (2008) continues, is especially dire in states like Kerala, where targeting has led to the economic collapse of a system that was most effective under universality. Proving the counterargument, i.e. the continued effectiveness of the PDS in the only state – Tamil Nadu – where the scheme remained universal, Swaminathan's work draws causal relations between the introduction of targeting and the overall deterioration of the PDS.

Second, against the backdrop of such deterioration, a substantial body of research explores the effectiveness of PDS reform. Reetika Khera and coauthors, including Delhi School of Economics welfare economist Jean Drèze, have worked extensively on the topic. In a paper published on India's *Economic and Political Weekly*, Khera (2011b) matches official figures on PDS offtake with NSS data on household consumption, showing that while a number of states were still 'languishing', several were 'reviving' the PDS through reforms. Offtake in such states, which involved both demand-related reforms (such as lowering commodity prices) and supply-related ones such as expanded coverage, has improved through the years, offering solid evidence that PDS reform can successfully address the issues induced by targeting. In a subsequent study, Drèze and Khera (2015) showed that PDS leakages were showing 'clear signs of improvement', especially in states undertaking 'bold reforms' on the sides of commodity prices and coverage.

Based on the by-then latest NSS data, such results on PDS reforms were however not echoed by the policies that the Economic Survey, an annual document of India's Ministry of Finance, suggested in 2015. In the chapter titled 'Wiping Every Tear from Every Eye: The JAM Trinity Number Solution', the Ministry started off by listing three maladies associated to in-kind subsidies. These were identified as how (1) price subsidies are often regressive, (2) price subsidies can distort markets in ways that ultimately hurt the poor and (3) leakages seriously undermine the effectiveness of product subsidies. Drawing examples from the PDS and other anti-poverty programmes in the nation, the chapter then compares the PDS to a technology-enabled alternative: direct benefit transfers, to be made directly on users' account through the Jan Dhan Yojana, Aadhaar and Mobile numbers (JAM) trinity solution. Such a solution combines Aadhaar's biometric identification with the Pradhan Mantri Jan Dhan Yojana (abbreviated as Jan Dhan), a central government's programme aimed at creating zero-balance bank accounts for low-income people, with mobile numbers; seeding bank accounts (in a way similar to what was done for ration cards) with Aadhaar, the programme would afford replacing PDS food rations with equivalent value transferred in cash, bypassing leakage and reaching the Nirvana that the national anti-poverty system long struggled for. As the chapter's conclusion notes,

> It will be a Nirvana for two reasons: the poor will be protected and provided for; and many prices in India will be liberated to perform their role of efficiently allocating resources in the economy and boosting long run growth. Even as it focuses on second generation and third generation reforms in factor markets, India will then be able to complete the basic first generation of economic reforms. (Government of India, 2015: 65)

But if technology-based, JAM trinity enabled cash transfers are meant to bring Nirvana to all, why would Adeela express such a strong preference for food? As our conversation proceeded, we started seeing the heart of the problems she shared with us. Similarly to Ankita, there is one core aspect she values in the PDS: having food rations on her family table, with the material security that – whatever happens with food quantities, cuts or delays – food will be there every month to be eaten. Adeela's problem was not, as it surfaced in other interviews across the state, the risk that men in the house would access cash to get other things than food and essential items. It was the fact that a prospected reality of cash in lieu of rations threw her in uncertainty, related to what would happen with the rations she needed so direly. The fact that her argument was echoed by our respondents – the large majority of whom were averse to cash transfers – illuminates a puzzling gap, in which the same people who, like Ankita, display enthusiasm for ABBA, at the same time are in fear of a shift to cash transfers, which Aadhaar and the JAM trinity of which it is an integral part were set to enable (Government of India, 2015).

Rejection of the cash transfer option from PDS users has been widely documented in the literature. In a study published in 2012, Raghav Puri – currently a Postdoctoral Associate at the Tata-Cornell Institute – illustrated the 'remarkable revival' of the PDS in the state of Chhattisgarh; partly as a result of this, people expressed a 'loud no to cash', with 93% of the surveyed people preferring the PDS to a system involving cash transfers (Puri, 2012: 21). Aggarwal (2011) similarly finds that surveyed people in Orissa, a state that was often associated with weaknesses in the PDS, express strong preference for food grains over cash, also as a result of the improvements that the state has introduced in the PDS after the targeting reforms. With Khera (2014) accounting for a survey of over 1200 rural households in nine Indian states, presented with the cash-or-food question, the overall finding spoke for itself; not only was preference for food overwhelming (67.2% of the households surveyed across all states), but it also was strongly associated with the benefits experienced under the current system, with topping records of preference for food in states classified as 'functioning' or 'reviving' according to their PDS performance (Khera, 2014: 116).

But it is here that the conundrum emerges. Not only does Adeela not realise the transition of PDS to cash transfers that Aadhaar is associated to, a policy which she fears, but also Ankita does not know, and for that matter, *none* of the persons we spoke to in our Karnataka fieldwork brought up the issue of Aadhaar being linked to such a policy. *Opacity* is, in this perspective, not a construct that details the technology's blackboxing of processes crucial to people's lives; it indicates the effect of people's inability to ask questions on technologies that, like Aadhaar and the JAM trinity, are associated to the intent of replacing anti-poverty systems that users have long availed. What is harmful for PDS users is not knowing – and more fundamentally, not being in the position to ask – what will change in the social protection system with Aadhaar's introduction, and not being able to react in the event that unwanted changes, such as a move to cash transfers, are implemented.

Opacity Made Stronger: Information-Erasing Technologies

A key aspect in Adeela's story is that, while she does not know of Aadhaar's functionality towards a move from the PDS to cash transfers, nobody is actively hiding this information from her. A possible transition to cash transfers is widely discussed in the media, and at the time of the interview, the cash-based system had already been piloted in Chandigarh, Puducherry and urban areas of Dadra and Nagar Haveli

(Hindustan Times, 2017). The system is not designed to hide this information from her; it does, however, put her in a position of either registering with Aadhaar or being cut off from food rations, an *aut aut* situation that does not encourage her to ask what Aadhaar registration entails. Like other PDS recipients in states using ABBA, she is not in the position to ask questions; while she has no information on the link of the PDS to cash transfers, she is not well-placed to investigate it.

A different situation occurs when it is *the artefact*, for how it is constructed, to erase information that is key to its users. Opacity alone, for how we defined it, does not entail an act of conscious omission or deletion of information. Below I discuss, however, what happens when information for digital ID users is actively edited out. A pioneering humanitarian blockchain project in Jordan and a subsidy scheme in Colombia both offer poignant instances of information-erasing systems.

Blockchain-for-Refugees: The Information-Erasing Machine

The case of refugee registration by the UNHCR, seen in Chapter 4, carries a wider message; the uptake of biometric technologies in the humanitarian sector is growing fast, embracing the logic of linking low-resource, often undocumented beneficiaries to data about them. The work of Margie Cheesman, a Lecturer in Digital Humanities at King's College London, offers an unprecedented sighting of the human consequences of this uptake. A digital anthropologist focusing on humanitarian technologies, Cheesman authored a PhD thesis titled 'Infrastructure Justice and Humanitarianism: Blockchain Promises in Practice', widely recognised as the first work ever to ethnographically examine the use of blockchain applications in the humanitarian sector.

Cheesman's thesis (2022a) opens with a colourful illustration of the hype, and also the corresponding downplay and meme-fication, around blockchains in the humanitarian context. The core feature of *blockchains*, she notes, is that of being decentralised digital infrastructures; rather than impinging on a central point of control, they distribute authority through a network of nodes, each of which is supposed to maintain a continuously updated ledger. Cheesman's (2022a) analysis starts by laying out the theoretical bases of the blockchain-for-development ideology; with its pliability and adaptability to human needs, the decentralised infrastructure of blockchain can fix the institutional failures which many centralised systems of development governance have illustrated. Blockchain arises, to use Heeks' (2002) terminology, as a route to fixing 'design-reality gaps'. By decentering development project architectures, the specific needs of different beneficiaries can be effectively incorporated into system design. This principle is the root of many imagined benefits, to the point of promising refugees a 'self-sovereign identity' mediated by blockchains (Cheesman, 2022a, 2022b).

Cheesman narrates, however, the lived experience of people at the receiving end of the innovative blockchain-for-development effort. Her ethnography centres on a pilot project collaboratively led by two UN agencies, pseudonymised as UN Basic Assistance (BASS) and UN Gender (GEN). Based in Jordan, the pilot consisted in using a shared blockchain to deliver and monitor the activities of the two agencies in the refugee camps of Za'atari and Azraq; not only would the two organisations get to use a shared blockchain, but this would optimise important practices in aid delivery. Chief among these practices, the project entailed switching how GEN paid the recipients of its cash-for-work programme, from a system of cash envelops to one in which beneficiaries, refugee women with work contracts of usually 3–6 months, would access their salaries through biometric iris scans.

Cash-for-work programmes are increasingly popular in systems of assistance to refugees. The logic is that vulnerable people, including forcedly resettled refugees and displaced persons, instead of receiving in-kind goods or donations would engage in work activities, regaining financial independence and an active social role as workers. *Digital wallets*, meaning financial applications allowing digital storage and withdrawal (and, in some cases, payment) of money, are often associated to cash-for-work programmes. By creating a virtual space for workers to receive their salaries, they provide the technological basis for the financial independence of structurally vulnerable subjects. Payments enabled by digital wallets, notes the UNHCR (2022), are being seen as a 'pathway towards financial inclusion', on which refugees can actively build to regain the self-sustenance that displacement took from them.

Women in the cash-for-work programme studied by Cheesman (2022a) were refugees displaced by the Syrian civil war, employed in activities ranging from cleaning to weaving, embroidery, teaching and childcare. Their salary payments, usually of 1–2 Jordanian dinars per hour, played an important part in the domestic economy of workers. At its heart, the UN blockchain project reconfigured the payment of salaries; before it, they received cash envelopes with stipends, a practice deemed as possibly risky by part of the social protection literature due to the difficulty of tracing cash (cf. Devereux & Vincent, 2010; Aker et al., 2016). With blockchain, payments would be made biometrically safe; the programme brought to the installation of EyePay, an iris scanning machine, in the camp's supermarket, where cash-for-work recipients would collect their salaries through iris-based authentication. At the moment of salary collection, women would go to the supermarket to perform the iris scan; uniquely identifying the user, the EyePay machine would generate a cryptographic key for them, associated to their digital wallet. The machine would then print out a receipt indicating the cash due, receipt that women would then bring to the supermarket cashier to collect their salaries.

The project offers a clear illumination of the blockchain-for-development ideology, where 'unsafe' cash envelops are replaced by a more reliable biometric architecture and digital wallets offer women a tool to independently manage their money. The reality narrated in Cheesman's ethnography (2022a, 2022c) is, however, very different. In an interview with five seamstresses in the Azraq camp, workers note how the artefact, i.e. the paper receipt, effectively hides key information instead of providing it. On the one hand, cash envelopes offered tangible clarity of what had been paid, and a humanitarian worker would be present in person in case any mistake should be corrected. On the other, the new system is not reported to do so. 'We never know what each salary payment is for', told a worker to Cheesman (2022c), and another one elaborated: 'We wish there was clear information somewhere telling us how much we received based on how many days of work, and when we received it, and how much we withdrew and if there's anything left'.

Cheesman's ethnography clearly illuminates the core of the problem. Women expanded widely on the importance of paper receipts; they would preserve them carefully, hiding them into their bras during the working day and folding the paper to prevent ink from going off. They would only take them out when having to withdraw or discuss their salaries. Written in English and with Roman numbers, receipts were unintelligible for most recipients, who often relied on collective accounting practices to interpret the information. Similarly to how Adeela, Ankita and others saw the physical, secure materiality of food rations as the most important thing, for refugees in Cheesman's study it is cash that counts. When they speak about 'receiving their salary', she notes, they mean cash in hand, not the amount stored in their digital wallet. Central aspects of the system, such as the cryptographic authentication key or the blockchain's distributed ledger, were invisible to them and beyond their awareness.

But the artefact, the paper receipt, displays a clear limitation. Women workers spoke diffusely about the information missing from it; specifically, information of what each salary payment was for, which days of work were being paid and when each salary payment was being received. All information that, Cheesman (2022c) continues to note, was available in the system based on cash envelopes, which also offered a much more direct route to the one item – cash in hand – that mattered most directly to them. On the one hand, in this project the blockchain did exactly what it promised to; it made salary payments secure, guaranteeing the traceability that UN donors – needing to access clear records on transparency – place at the top of their interests. But on the other, it does a new thing; with its produced artefact, the receipt, it hides information that is key to recipients, factually turning itself into an *information-erasing machine*. The programme perpetuates informational injustice by deleting key information for digital ID users, information that old technologies, with their cash-based architectures, provided in ways that worked well for recipients.

Reflecting on the experience of aid workers and refugees suddenly faced with blockchain, Cheesman coins the notion of *infrastructural justice*, which I enter into dialogue with in Chapter 8. Infrastructural justice, Cheesman (2022a: 20) argues, is closely predicated on three aspects: subjectivities (the range of affective, embodied and culturally situated rationales and responses connected with blockchain), timescapes (the temporal and spatial regimes and routines blockchain disrupts and co-creates) and materialities (the material practices, processes and architectures that are established and maintained with blockchain infrastructure). As the notion of infrastructural justice participates in the theoretical landscape which this book seeks to enrich, Chapter 8 puts it in conversation with the taxonomy of injustices that this book develops. It is to be noted, for now, that informational injustices – caused by information-erasing practices that directly hurt refugee women – are a substantial component of the cash-for-work programme described here, and actively contribute to the way in which infrastructures participate in injustice generation.

Information-Erasing Databases: Ingreso Solidario in Colombia

The paper receipts studied by Cheesman (2022a, 2022c) are an artefact that is physically in the hand of beneficiaries of a cash-for-work programme, and hides key information right in the act of mediating their access to the system. Informational injustice, however, is not only a problem of artefacts held directly in the hands of users. On the contrary, research on computerised welfare programmes shows that a key *locus* of deletion of information, and especially of the way particular pieces of information were produced, may be the very databases where user information is held. It is in such databases, stored in the core of digital identity platforms, that lie the informational basis upon which many decisions are taken, first and foremost the assignation of particular welfare entitlements to users.

This matter takes the focus to Colombia, whose social protection system is studied by Joan López Solano, currently a PhD candidate in Law, Technology, and Society (TILT) at Tilburg Law School. At our first, virtual encounter in the making of the book *COVID-19 from the Margins: Pandemic Invisibilities, Policies and Resistance in the Datafied Society* (Milan et al., 2021), López worked for the Fundación Karisma, a Colombian civil society organisation engaging digital technologies to protect human rights and support social justice. Through his work with recipients of technology-mediated social protection programmes, López conducted extensive work detailing people's lived experience of computerised social welfare. His contribution to *COVID-19 from the Margins* was centred on *Ingreso Solidario* (Solidarity Income), a cash transfer scheme devised by the Colombian government to respond to income losses suffered by families in poverty during lockdown.

Colombia operates a scoring-based system of vulnerability evaluation on households. Since 1995, the government operates the *System of Identification of Social Program Beneficiaries (Sisbén)*, a unified household vulnerability index made to identify poor households for enabling them to access social protection schemes. Different from the Indian system that classified households as extremely poor, below-poverty line or above-poverty-line, Sisbén is based on the classification of four groups: A (extreme poverty), B (poverty), C (vulnerability) and D (not poor or vulnerable), to which each household is assigned based on socioeconomic variables. Calculation of the index is based on three components: a socioeconomic survey to collect data on households; a welfare measure to assess vulnerabilities; and software to perform household level calculations (ILO, 2017).

Two main characteristics of Sisbén are notable. First is its wide recognition as a scoring system that shifted Colombia's social protection policy from demand- to supply-oriented, actioning the logic of information for development championed by the World Bank (1998). The Sisbén database, which in 2014 held information on over 70% of the total national population, was derived by the combination of the household survey with the welfare index, which allegedly afforded prompt identification of needy families (ILO, 2017) and assignation of entitlements to them. But the system was upgraded over time; while its first version, Sisbén I, was designed as a proxy for resources and income, its present version, Sisbén IV, incorporates 24 variables across dimensions of health, education, housing and vulnerability. Enforced by the National Planning Department (NPD) across all municipalities, Sisbén IV is used in the selection of beneficiaries of more than 19 social protection programmes, ranging from poverty alleviation to housing and healthcare (López, 2022: 2).

Second is a crucial, distinctive characteristic of the system, especially illuminated by López (2022) in his report *Data for dignity: Requirements for the implementation of data systems for social programs in Colombia*. By design, Sisbén is meant to be a system that uses inputs *from households* to determine their vulnerability scores; there are both sweeping surveys, aimed at building the initial database, and on-demand surveys, for households needing to add or remove members or request a new assessment of their vulnerabilities (ILO, 2017: 3). Similar to other versions, Sisbén IV does not only triangulate data with social context, but also with the affordances of each family; for instance, a household based in an area of poor public service delivery is meant to get a lower score (meaning, higher vulnerability) than a family with the same descriptors and number of members, but based in a better served area. Its peculiarity, making it different from other versions of Sisbén, is the interoperability of the social register with multiple public and private databases. The logic underlying interoperability is that of identifying the needy so to distribute scarce resources in an optimal way. Over the years, Sisbén has been recognised as the basis for assigning benefits associated to better nutrition, health and education outcomes for beneficiaries (Baez & Camacho, 2011; Melguizo et al., 2016).

Yet, López (2022) reveals the presence of fundamental issues with Sisbén IV. The system, he shows (2022: 3), reproduces the conception of social security as a form of charity rather than a fundamental right. Its underlying epistemology mirrors the limitations of targeted social protection, which, as noted in Devereux (2016), risks depriving large sections of the population from basic rights. As detailed through the report from López (2022), the system focuses on 'identifying' people worthy of accessing subsidies, as opposed to ensuring a minimum standard of living through the community'. In such a system, the burden of proof for a needy status shifts on the beneficiary. It is here that the opacity of subsidy assignation, granting no detail on why a given benefit is or is not assigned to a household, puts vulnerable people into especially high predicament (López, 2022: 8–9).

Against this backdrop, the COVID-19 pandemic has marked a route to welfare delivery that crystallises the issues emerged with Sisbén IV. Just into the first lockdown of 2020, animated by the decision to support families unreached by existing subsidies, the Colombian government has introduced Ingreso Solidario – an unconditional cash transfer scheme built for families in poverty, who do not receive any other cash transfer from the state. With its gendered distribution (63% of recipients were women) and the speed of its creation, over just two weeks during the first pandemic lockdown in March 2020, the programme was strongly advocated by the Better Than Cash Alliance (2021):

> The Government of Colombia's experience in fostering public-private collaboration is an inspiring example of how digital payments can be rapidly delivered across multiple channels. The case of Ingreso Solidario demonstrates that designing programs responsibly – aligning with the UN Principles for Responsible Digital Payments – can increase trust in digital financial services, leading to greater volumes of digital transactions, and increased use of digital savings and e-commerce services. (Better Than Cash Alliance, 2021: n.p.)

The theme of 'responsibility' of digital payments is echoed beyond the Better Than Cash Alliance's assessment. Ana Maria Prieto, Deputy Director for Market Development at Colombia's Ministry of Finance, noted how the programme would transcend COVID-19 emergency response, to 'leap-frog financial inclusion and lay a new foundation for G2P payments in the country'. Her assertion was based on the presence of a Consolidated National Social Registry, including all the beneficiaries of social programmes. She noted how government-to-person (G2P) payments mandated since 2017, along with a friendly regulatory environment, would enable transformational results beyond the short term (Prieto, 2021). Associated to a reduction in poverty estimated as 0.8% points (Cuesta & Pico, 2020), the programme displayed grounds for obtaining large international favour in the face of pandemic hardship.

But what this story does not show, notes López (2021a), is the process of selection of beneficiaries that, in multiple ways, have been incorporated in Ingreso Solidario. The government's recount is that of a straightforward process. In an effort to search for the unreached, it used data on vulnerability scores from Sisbén IV, combined with other databases that could ensure solid triangulations. Yet the nature of such 'other databases' remains uncertain. The programme, notes López (2021a), has built a Master Information Database in which the NPD 'mixed different administrative records using data collected for different purposes, managed by private and public actors'. The mystery remains on the exact nature of the data collection involved. In this experiment, as López (2021a) recounts it, even databases of prisons and the Forensic Medicine Institute were used to try to deduplicate records. Differently from Sisbén, where households play an active role in their vulnerability assessment, Ingreso Solidario has deprived assessed people of their agency in determining or contesting decisions on them.

The problem has at least two more components. First, the ambition praised by the World Bank is largely based on the determination to 'reach the unbanked', identifying people unreached by existing social payment strategies. Beneficiaries who already have a bank account, note Velez (2020), were paid into that account, while those who did not have one received messages on their mobile phones, instructing them to open a digital account with a bank, public or private. On the one hand this process is epitomic of the public–private connections lying behind the system's data handling processes; this presents one more difference with the idea of using sources that were known to the assessed people. On the other, as Velez (2020: 22) observe, the issue lies at the back-end: such payment digitalisation raises questions about ethics in handling the data of the applicants, the transparency in reception by the real beneficiaries and possible hidden profit for the financial system.

Secondly, beyond obscurities in beneficiary selection, the cross-checking practices of the scheme have generated issues of outcome. López (2021a) details how, when the first dataset from Ingreso Solidario was published, multiple complaints on ghost beneficiaries were raised; many from the public reported names of dead people, also identifying non-existing and expired ID cards. In response to this, the NPD dismantled the database and gave it to the national registry for deduplication. This resulted in 17,000 records found to have inconsistencies, generating vivid debate on the extent to which a cross-checking strategy would pay off (El Tiempo, 2020). Using vast amounts of data within its reach, the government has used as much data as possible to find the fewest beneficiaries, fighting a war to inclusion errors that begs the question, López (2021a) notes, on what happened to those needy households that the system has not seen or calculated.

Cheesman's work with refugees had shown what happens when artefacts in the hands of people are data-erasing, hiding pieces of information that are important to the user. With his work on Sisbén IV and Ingreso Solidario, López shows that the problem can be at the back-end; relevant data can be erased at the moment of their assemblage, obscuring how their cross-checking is conducted and validated. With a database that deprives assessed people of their agency, his work shows that informational injustice can go well beyond papers-in-hand. It can effectively lie in the way individuals are profiled, to be then represented and treated through architectures based on digital ID. Thanks to his concept of *data for dignity*, where data are framed as a route to restore the epistemic violence perpetrated on people, we will explore different forms of fair ID practice in Chapter 8.

On Information Orders and Injustice

The close of this chapter takes us back to the notion of information orders as elaborated in Srinivasan (2022). Her modified notion of an information order leads us to reflect on how channels, forms and agents shape not only information, but also the production and experience of informational injustice as it has been illustrated through the chapter. This point is advanced below.

With her ethnography of the *political lives* of information, Srinivasan (2022) invites the reader to pay attention to how information orders shape the way information is produced, received and leveraged by the actors involved with it. She modifies Bayly's (2000) notion of an information order as a socio-technological formation to include the artefacts (information systems, laws, documents, procedures) that shape information, as well as the physical and social infrastructure through which it is circulated. Not only did her modified concept of an information order afford her to study information through rigorous ethnographic work, but it also put her in the position to challenge essentialist understandings of information, based on economistic models which do not account for people's lived realities. Her study of information orders thereby illuminates both the production of information, and the politics connected to its leverage from actors whose power is specifically produced within contexts of interest.

This sociotechnical, materialised vision of information helps us understand informational injustice in the forms in which this chapter has studied it. In my study of the PDS in Karnataka, what counts most for recipients is tangible, understandable information on rations; a ration shop blackboard showing entitlements by category, a door sign indicating distribution days and variations in them.

Inevitably, this is information on the one matter that recipients value most, meaning the food rations to be availed on the family table every month. It is not back-end information on Aadhaar, its data handling techniques, or the management of biometric credentials that users register with it; those data escape the recipients' worldviews, as users are put in the *aut aut* condition of either doing Aadhaar seeding, which requires enrolment, or not being at all enabled to receive food rations. Information of primary importance, relative to monthly food rations, relegates other information points as secondary; this is also true for information on Aadhaar's functionality for a shift to cash transfers, whose possibility is met with concern. Injustice here, it is worth reiterating, does not come in any way from manipulation of information; it comes from putting the recipient in the position of asking no questions, as their need for rations is satisfied only upon Aadhaar enrolment and ration card seeding.

In the case of the humanitarian blockchain project (Cheesman 2022a, 2022c), the situation is different. Here we are confronted with an artefact, the biometrically produced EyePay receipt, which embodies a specific system of interests; even with their material tangibility, the receipts do little to replace the cash envelopes that were most important to refugee workers. Here, defined as a data-erasing artefact, EyePay receipts are to be seen in their sociotechnical context: what women workers preserve in them – hiding them in their bras, folding them to prevent ink from going off – are Roman numerals that, as Cheesman (2022a) observes, women workers approach with a logic of *faith*, as accounting practices are built on reliance on the few community members that can interpret them. But faith alone does not replace what the artefact takes away; that is precious information on what each payment is for, to how many worked days it corresponds and when the money was received, all information that was granted in the system based on cash envelopes. In this case, Srinivasan's notion of information order illuminates a specific, embodied system of interests; for UN donors, whose chief interest lies in the accountability of payments, biometrically produced receipts are an optimal tool to transparently keep track of transactions. It is for refugee workers, whose need is to receive cash in hand, that the same technology is suboptimal. Their interests are however lost in the design of the artefact, which reflects the priorities of donors and relegate theirs on a secondary level. Similarly to PDS recipients, they value material entitlements, but what they value is nowhere to be seen in the technology built for them.

Finally, we have the case of the Colombian social protection scheme Ingreso Solidario. This does not entail an artefact in the hands of people; what the NPD builds is a back-end database, which displays the data of beneficiaries to whom a much needed, emergency cash transfer is granted. But it is exactly the production of the database that illuminates the importance of the politics of information. With an obscure combination of existing databases used in the selection of beneficiaries, the system overturns the logic of people-led inputs; rather than households partici-pating in the input of their own data for vulnerability assessment, it is the Master Information Database that develops scores based on sources of which beneficiaries are not completely aware. The selection of beneficiaries, which comes as a result, is the byproduct of a process where relevant data can be erased at the moment of assemblage, which takes place in a back-end space that is invisible to users. The politics of the artefact is depicted as an effective mechanism of 'searching for the poor'; what is lost, however, is the agency that beneficiaries used to have in vulnerability assessment, leading to unquantifiable exclusion errors and avoidance of the political discussion on beneficiary selection (López, 2021a).

At least three considerations should be made on how Srinivasan's modified notion of an information order contributes to the point made here. First, Sriniva-san's notion presents a fundamental difference with Bayly's; on the one hand, Bayly saw information orders as more or less autonomous formations, generated in a precise historical moment and with specific political traits. For Srinivasan, the genesis of information orders is different – her modified concept sees information orders as directly shaped by the channels, forms and agents through which infor-mation is produced and leveraged. Without those channels, interactions and especially power dynamics (shaping the *political lives* of information) our study of the 'backrooms' of informational injustice (cf. Parmiggiani et al., 2022) would not have been possible. It is thanks to Srinivasan's modified concept that we can see information in its sociotechnical space. It is this vision that attaches deep meaning to a sign placed on a ration shop's door, a paper receipt in the hand of a refugee worker, a text message inviting an indigent person to open a bank account. The sociotechnical production, and its political leverage, of information in all three cases has helped us understand the genesis of injustice, within hidden layers that lie behind the front-end visibility of the exclusions that Chapter 4 has been dealing with.

Second, Srinivasan explicitly invites us to pay attention to the material artefact in which information is produced, codified and transmitted. Our journey through informational injustice has revealed how the artefact, and its positioning inside or outside the spaces where the user encounters the provider, speaks of the fairness of the informational experience they live. The refugee who authenticates with EyePay has a receipt in her hand when she collects her salary; but that same artefact that gives her the affordance of money collection takes from her the information she needs, on the days paid to her and the exact times of payment. On the walls of ration shops are boards with category entitlements, on the door the informational signs on opening times: this is the information that PDS users reach out to, rather than potential for a cash transfer transition. The text message inviting the person to open a bank account reports the outcome of the beneficiary selection, but says nothing on the data used to make their decision, or on the ethics of the government accessing such data. It is thanks to Srinivasan's concept that the materiality of artefacts has shaped our discussion of informational injustice.

Finally, we cannot ignore the relations between the concept of information orders and that of *politics of anti-poverty artefacts* that the previous chapter has explored. The concepts act in conversation with each other; one illuminates the genesis of information, positioning it in its sociotechnical dimension and arguing that not only does information have politics, but it goes through *political lives* that shape it into what it is. The other notes how artefacts reflect development policies with precise historicities, targets and choices made on the basis of their social context. In Chapter 8, we will see that these concepts, while important in theorising digital ID, do not act on their own, but are part of a broader conceptual armoury, built to provide a situated vision of digital identity which affords us to throw more light on what happens to the user when their identity becomes digital.

Summary

After having confronted the straight-up issue of exclusions, this chapter has taken us through the more hidden layers of informational injustices to which people are subjected. Srinivasan's modified notion of an information order, along with the return of the anti-poverty artefacts which we had seen in Chapter 4, has allowed us to understand what happens when key information is hidden from digital ID users. Combined with legal injustice, this notion contributes to us painting a more complete response to the question on *what happens to the user* as a result of digital identification. The answer, however, would be incomplete if we did not venture into the heart of digital ID design, which we do with the notion of design-related injustice.

6

DESIGN-RELATED INJUSTICE

Chapter 5 has illustrated how informational injustice takes place, through the obscuration of essential pieces of information produced through digital identity systems on users and their entitlements. Diversely from legal injustice, which people encounter in the direct form of denial of legal rights and entitlements, informational injustice acts at a more hidden layer, beyond the interface where the user meets the provider.

But layers of digital ID-induced injustice can be even deeper. Rather than playing out as unintended and unplanned, injustice can be scripted directly into how the technology is designed, in silent ways that the subject, encountering the technology at its front-end interface, is not positioned to imagine. It is here that a design dimension, having to do with the conceptual plan lying behind the technology, becomes relevant to the production of more or less fair outcomes.

What's in Design?

As my co-researcher and I approached the ration shops in Karnataka, ration dealers took us through the novel Aadhaar-based interface, which after ABBA's inception recognised people's fingerprints and matched them to a ration card with entitlements (*The Economic Times*, 2016). The ration dealer's laptop, supplied by the state, is connected to a fingerprint reader, through which people would be, or not be, recognised as entitled users. The scale on which goods were weighed was separate from the architecture of user recognition. As we will learn in this chapter, this was different from the system I had observed during previous ration shop visits in 2014 (Masiero & Prakash, 2015). Back then, a pre-Aadhaar system was established at the state level by the Secretary of Food, Civil Supplies and Consumer Affairs. In that early version, the point-of-sale machine was not separate from the weighing scale; it was incorporated in it, so that weighing and disbursement of goods would take place as one operation. As this chapter will note, such an architecture was crucial to the planned disbursement of correct quotas of goods to people.

In showing us the laptop and biometric reader, one ration dealer that we spoke to was especially explicit in clarifying how the system protected the user from issues of leakage from the PDS: 'no more corruption is possible', he told us comparing ABBA with the non-digital system previously in place. His shop assistant reinforced the message; with ABBA it is not possible to divert PDS goods to the market, because the system's architecture makes transactions conditional to the biometric authentication of the individual. This architecture seeks to dismantle the guilt allegation long associated to ration dealers. As noted in Chapter 4, ABBA is scripted to verify user entitlements, but also to prevent ration dealers from subtracting goods from the system. As shown in Figure 6.1, the system puts Aadhaar into communication with the Food Corporation of India's (FCI) database of PDS users; this monitors both the users, who are called to prove their identity and entitlement, and the ration dealers, who are held to account for all transactions made in their shops.

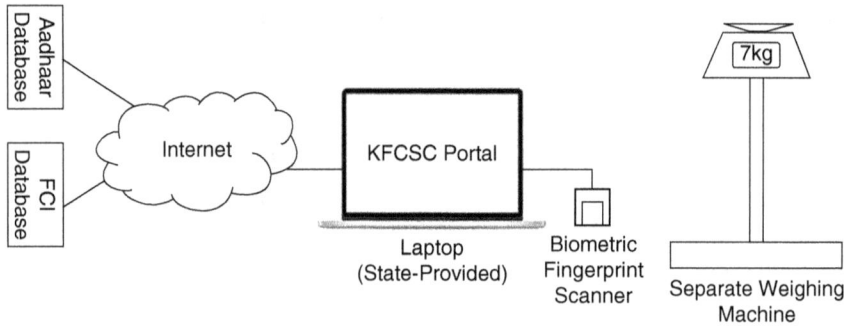

Figure 6.1   Aadhaar-Based PDS Architecture in Karnataka
Source: Masiero and Das (2019: 921).

Seeing the system from the ration dealers' perspective allows us to delve into design, with which I mean *the conceptual plan* that makes an information system what it is. Figure 6.1 illustrates the design of Karnataka's Aadhaar-based PDS architecture, and reveals key relations in it. First, the shop interface is underpinned by a user recognition system, which subordinates disbursement of goods to correct authentication of customers. Second, this nexus is predicated on the combination of two databases, those of Aadhaar and the FCI, which makes Aadhaar seeding of ration cards a prerequisite for user authentication. Third, the same nexus involves also the ration dealer, who needs to prove Aadhaar-based sale of goods to collect commodities from their local FCI warehouse each month. These features make the ration dealer accountable to the user, with a logic that seeks to liberate them from the allegations of guilt for corruption (Khera, 2011b) long associated to their figure.

A design perspective, illuminating the plan behind the system, shows us aspects of ABBA in Karnataka that we could not have grasped otherwise. These involve both the mechanisms of functioning of the Aadhaar-based PDS, and the problems that these can generate. On the one hand, the ration dealer is happy with a newly obtained accountability, and several made that clear in conversation with us; but on the other, one key mechanism with which diversion happens is not directly affected in ABBA, due to the separation of the weighing scale from the point-of-sale machine. As noted in field research conducted by Hundal et al. (2020), this feature disconnects the measurement of commodities sold from the act of disbursement; ration dealers can sell lower quantities to users, simply by altering the quotas they effectively disburse (cf. Masiero & Prakash, 2015, 2019; Prakash & Masiero, 2015).

A second interpretation of the system's design concerns the way data are aggregated under ABBA. Data aggregation, matching Aadhaar credentials with household ration card details, is epitomic of how ABBA subordinates subsidy assignation to Aadhaar-based registration and authentication. The design of the system reifies the conditionality discussed in Chapter 4; a fundamental right, such as the right to food, is subordinated to correct authentication of the user through the system's interface. In other words, the 'voluntary' nature of ABBA is denied by its own architecture; the subordination of service provision to correct authentication is written into the artefact, and cannot be separated from its functioning.

But it is a third interpretation that brings us directly into the heart of the problem. We introduced it in Chapter 4, with the notion of legal injustice; by making access to the PDS conditional to Aadhaar-based recognition, technology is not designed to address the plea of all those being left out, finding themselves in the condition of being entitled to the system but not recognised as such. The system's design carries a precise purpose; that of enforcing exclusion of those who do not comply, reifying the mechanism that turns the non-compliant in a non-subject of the same rights. The 'excluded needy' spoken about by Swaminathan (2002), who are entitled to PDS subsidies but not recognised as such, become a second-order problem as reified by technology design; this is no incidental feature of the artefact, but a matter at the very heart of how it was designed.

As we visited ration shops across Karnataka, users for whom ABBA worked well gave us enthusiastic accounts of the system, at all points being able to collect their entitlements. But the same architecture is designed to produce the exclusions that stem from it, exactly because *inclusion errors*, against which the technology cautions, are prioritised in front of exclusion errors. This marks the leap between what research calls 'unintended' consequences of technology, and the notion of design-related injustice, meaning *injustice perpetrated through technology design, based on the technical features of digital identity systems.* It is this form of injustice, directly designed into digital ID systems, that this chapter engages.

Not a Dark Side

The book 'Design Justice: Community-Led Practices to Build the Worlds We Need', authored by MIT researcher Sasha Costanza-Chock (2020), is a strong guiding light in understanding design justice. The book narrates the very genesis of the term 'Design Justice', stemming from a community committed to sharing 'ideas and strategies' for how to create a more just, creative, and collaborative world (Costanza-Chock, 2020: 1). Born during the Allied Media Conference (AMC) in 2015, the Design Justice Network puts together 'designers, artists, technologists, and community organizers' who participated in the workshop 'Generating Shared Principles for Design Justice', through which a living document of principles was generated (Costanza-Chock, 2020: 1–6). Rather than a static theoretical lens, design justice offers a perspective in continuous evolution, aiming to capture both the injustice implicit in technology design and the role of design in justice restoration.

The heart of Costanza-Chock's (2020) exposition of design justice lies in the universalism of design principles, and in how this exerts the power to marginalise and ultimately erase large groups of people. Writing from a queer and trans* lens, she starts the book narrating a security control situation in an US airport, in which her body's difference from the gender norm of female bodies results into her being flagged as a potential security threat. The cis-normative architecture of airport scanners, along with the discomfort and harm generated on flagged users, arise as a paradigm of design injustice; that is of the erasure of marginalised communities from the silent 'norm' inscribed in technology. Relying on Winner's (1980) notion that artefacts have politics, the book illuminates how 'cis-normativity is enforced at multiple levels of a traveller's interaction with airport security systems' (Costanza-Chock, 2020: 4).

Costanza-Chock (2020) puts forward design justice principles as a byproduct of the Design Justice Network, principles to be intended as collectively produced and developed. At the heart of design justice principles is the notion of intersectionality, seen, with Crenshaw (1991), as the intersection of multiple overlapping systems of oppression. Diverse forms of injustice, along lines from gender to race and disability, can be inscribed in technology. Costanza-Chock (2020) is clear on how the recognition of injustice feeds the making of design justice, which generatively plies design in the attempt to restore people's rights. In a way, the Design Justice Network offers a lens to 'build a better world' not through incremental improvement of existing technologies, but through recognition of the liberating power of design in countering the injustices that technologies can perpetrate.

Now, if ideas of design justice are espoused, the vision of a 'dark side' of technology is problematised at its basis. Diffused in Information Systems research, the vision of a 'dark side' is well exemplified by Tarafdar et al. (2015); the underlying idea is that technology is designed with beneficial purposes, and the harm it generates comes from unintended aberrations classifiable as 'side' effects (cf. Tarafdar et al., 2015; Salo et al., 2022). The argument of a design justice lens is different; far from being incidental, injustice is designed into the body of technology, which produces harmful outcomes exactly as a result of its planned design. In this way, the argument of Costanza-Chock (2020) reconnects to that of Winner (1980). If the artefact is designed to enforce unjust policies, it cannot be separated from the unfair decisions built in it. What is being looked at is not a 'side', but the very architecture of technology, purposefully planned and built.

Two key points should be made on the concept, *design-related injustice*, at the centre of this chapter. The first point relates to the concept's genesis; when we first developed the framework published in Masiero and Das (2019), our concept was milder, and connected to the notion of *design-reality gaps* theorised by Heeks (2002). Our fieldwork on the PDS revealed a gap between the world of designers, for whom the priority was to fight inclusion errors, and the world of respondents, whose overarching priority was food on their tables. For this reason, noting that ABBA tackled inclusion errors but did not intervene on exclusion, we had argued that design-related injustice stemmed from a technology that de-prioritised user needs. Having witnessed the hardship of exclusion errors on people, we had seen design-related injustice as a result of such mismatch in priorities.

Over time, however, our conceptualisation evolved. Through encounter with Costanza-Chock's (2020) work, one main reflection ensued; de-prioritisation of user needs is indeed a problem, but the deeper issue lies in how harmful outcomes are being inscribed directly in the technology's body. As Costanza-Chock displays, not only priorities, but exclusionary norms can be inscribed in technology design. The artefact that discriminates in virtue of the non-conformity of bodies goes beyond reflecting particular visions; it directly harms the person, putting them in situations of discomfort and potential peril.

The second point is the reason for sticking to the notion of *design-related* injustice, even after the eye-opening encounter with Costanza-Chock's (2020) *design justice* concept. On the one hand, Costanza-Chock shows how injustice is designed in technology, of which airport scanners, with their detrimental impact on non-conforming bodies, are a powerful illustration. On the other, my field notes point to an issue connected to this; injustice can be *related* to design, in the sense that it is produced by what design achieves in its interaction with people and the context within which they operate. This point is unpacked by the technologies of identification and tracing that we encounter below.

Biography of a Biometric Artefact

This book has, so far, immersed the reader in stories of injustice. So much so that it may run the risk to neglect what happens when technology participates in policies that aim to benefit recipients, for example by strengthening programmes as crucial as the food security schemes we have been talking about. To encounter one such case, however, we do not need to travel far away from the ration shops from which this chapter has started. We do, in fact, go back to ration shops in Karnataka, but with a story of computerised ration distribution dating back to the year 2014.

For context, ABBA had not been implemented in the ration shops back then. My co-researcher Amit Prakash and I sought to understand the changes in the state-level PDS made under B.A. Harish Gowda, who had back then recently ended his mandate as Karnataka's Secretary of Food, Civil Supplies and Consumer Affairs. It was under Gowda's administration that the Government of Karnataka, in timely correspondence with the approval of the NFSA, reformed the PDS through the *Anna Bhagya* scheme. Launched in 2013, it implemented a policy of strong subsidisation, where rice and wheat were to be sold at Rs. 1 per kg to the below-poverty-line (BPL) and distributed in varying quantities according to household size (see Table 6.1). There was, conversely, no subsidy planned for the above-poverty-line (APL), and an even stronger subsidisation – fixed at 29 kg of rice at Rs. 1 per kg per household – for the Antyodaya Anna Yojana (AAY).

Table 6.1 Entitlement to Food Grains under the Karnataka PDS (2014)

Status	Entitlement – Quotas (Per Size of Household)	Entitlement – Price
APL	None	None
BPL	1 member: 8 kg. 2 members: 16 kg. 3 or more members: 24 kg.	Rice: Rs. 1/- per kg. Wheat: Rs. 1/- per kg.
AAY	29 kg. (any household size)	

The historical context in which the Anna Bhagya scheme was implemented is key to the picture. As noted in Mooij (1994), Karnataka's PDS has a history of overall good functioning, with levels of offtake and utilisation comparatively high within the country. Yet, the state suffered from the same issues that the move to targeting brought across the nation; leakage of PDS items to non-entitled users was estimated at 46% in 2010, and ration dealers, already facing situations of unviability in the universal system, found themselves even more impoverished by the shrink in their customer base (Drèze & Khera, 2015). Against the backdrop of a PDS in crisis, by-then Secretary Harish Gowda embarked on a programme of computerisation – conducted in collaboration with the National Informatics Centre (NIC) Karnataka – that, while running a biometric PDS, differed substantially from the ABBA version implemented years later.

As we visited ration shops in three districts of Karnataka during August 2014, the user interface we encountered differed significantly from the Aadhaar-based one that was to come. Instead of a point-of-sale machine separated from a scale, what we met was a weighing *and* point-of-sale machine: users would have their fingerprints read by a device incorporated in the scale, where a display would report the name of the user and their entitlement. Produced by the Japanese company Essae Teraoka, the machine had speakers that, at the moment of disbursal, would announce the exact quantity sold by the dealer. This in an attempt to curb the fraud that, as noted by Hundal et al. (2020), can still take place in the case of an artefact that, like ABBA, disentangles user recognition from disbursement. One such machine is portrayed in Figure 6.2.

Figure 6.2 Point-of-Sale Machine, Karnataka, August 2014

In line with Sriraman (2018), an *artefact biography* approach is taken in order to understand the design rationale for the weighing-selling machine, which, at the time of our fieldwork, had been introduced in six of Karnataka's 29 districts. In a detailed, half-day conversation in August 2014, Gowda took us through the history of his efforts to digitise the PDS; computerisation of the scheme had started back in

2005, with the construction of a database of entitled PDS users. Such a database, popularly called *Ahara* ('food' in the local language), aimed at storing all details of ration card holders. A website, *ahara.kar.nic.in*, contained all such details, both for people's access to their own records and for public scrutiny. The public vigilance goal of Ahara was well-integrated with vigilance committees; groups of residents tasked with surveilling the operations of the ration shop, and whose members' names we often found affixed inside the shops' premises.

While well-functioning at the time that Gowda demonstrated it to us, Ahara followed, we found on the same day, a failed effort in the same direction. In 2009, the state's ration card management was outsourced to a private vendor. Rather than conducting proper checks of entitlements, the vendor resorted to issuing 'temporary ration cards', a document released to households while awaiting the due entitlement checks. As noted by the Justice Wadhwa Committee Report on Karnataka in 2010, the situation quickly went out of hand. In 2010, the number of ration cards in the state exceeded the number of resident households, leading to high leakage rates. With the entrance of a private actor in a high-stake process such as ration card management, the risk of outsourcing a crucial public service became evident, requiring the Gowda administration to take a different approach to entitlement determination.

Such an approach was radical. Rather than implementing a new door-to-door survey, the Secretariat required verification of ration cards' validity based on two possible routes: urban households had to provide a RR number (signifying a valid electricity connection) and rural households would provide a property identification number. All cardholders were requested to provide such details at centres, called photo-bio centres, set up across the state: in such centres a photo ID and fingerprints would be captured, in a pre-Aadhaar effort of biometric identification. It was this granular effort, rather than the centralised push exerted by Aadhaar, that generated the Ahara database and it was Ahara details that the weighing-selling machines would search to determine household entitlement.

But another, substantial difference emerges between this system and ABBA. Let us remember that, as already noted in Ramakumar (2011), a substantial amount of diversion does not happen in the ration shops. It occurs instead before goods even reach them, during transportation and storage in the FCI warehouses (known as *godowns*) from which ration dealers collect them. In Karnataka, biometric recognition in ration shops has been accompanied by the Financial and Accounting System (FIST): a system aimed at checking PDS transactions that occur in the godowns where commodities are stored and lifted from ration dealers. Operated by staff at Karnataka's godowns, FIST registered the amount of goods that come in from the FCI and private producers every month, and the amount that was lifted monthly by every ration dealer. The monthly amount of goods assigned to each ration shop is based on the theoretical requirement (determined through the number of house-holds registered with each shop and their poverty status), and the closing balance, meaning the commodities left in ration shops at the end of its month. In other words, without involvement of Aadhaar, FIST was built in such a way to perform a monthly accountability check of ration dealers. Figure 6.3 portrays a PDS godown.

Figure 6.3 PDS Godown, Karnataka, August 2014

It could be imagined, at this point, that the pre-Aadhaar digitisation of the PDS in Karnataka was an optimal instantiation of digitally enforced reforms. As we argued shortly after fieldwork (Prakash & Masiero, 2015), this was not the case. From the analysis conducted back then, at least two issues remained. First, quality control on the food items disbursed was problematic. In multiple occasions, we heard from PDS recipients feeding PDS rice to goats or chicken, due to the low quality of supply; we ourselves witnessed, during trips to rural districts, several instances of rice bags corrupted with mud and stones. Despite the presence of quality inspectors at the FCI level, quality control was witnessed as largely absent from the PDS, with the risk of affecting the distributional architecture that the new digital system was seeking to facilitate.

Secondly, incentives to diversion for ration dealers remained. As noted in Drèze and Khera (2015), the post-targeting situation remained tough for them. We encountered several dealers conducting other businesses than the ration shop, in pursuit of financial viability. We witnessed, at the same time, multiple forms of machine tampering on the ration dealers' side; the speakers, which were meant to announce the quantity of goods being sold, were muted in most shops we visited.

Senior officials interviewed at the time agreed that dealers can alter the weighing mechanism, refuse to provide bills to beneficiaries and engage in forms of tampering that still allowed diversion. In the face of all this, the problem raised in Khera (2011b: 1048) remains central; the financial unviability of ration shops can make corruption a requirement for survival, a problem that needs attention at the policy level rather than just at that of implementation.

But a point remains on the weighing-selling machine, later supplanted by ABBA. With all its limitations, and exclusions of users still persisting (Masiero & Prakash, 2015), the system championed by Gowda constituted an effort to *revamp* the PDS rather than dismantling it. At a time when the JAM trinity, with Aadhaar at its core, set to enable the transition from PDS to cash transfers (Government of India, 2015), the state of Karnataka proposed an artefact aimed at reinforcing the PDS, rather than easing its transition to a different scheme. As I argued at the time (Masiero, 2015b), the JAM trinity was not bound to dismantle the PDS: in the hands of leaders supportive of redistributional justice through a system based on subsidies, its core artefacts could effectively strengthen the food subsidy system.

Biometric Artefacts, Policing and Partnerships

As I read it *ex post*, the artefact's biography of the weighing-selling machines of Karnataka has two main messages. First, artefacts plied to an unwanted purpose (such as the cash transfers that many respondents fear) can be used to completely reorient their initial goal; that is to reinforce the subsidy scheme that caters to people, as Harish Gowda's scheme had energetically tried to do. Second, it is the policy inscribed into the design of technology that dictates its goals, placed in the user reality where encounters with providers are produced. To deepen this point, we move beyond social protection technologies to first examine a set of tools, referred to as *border technologies*, where the focus shifts from protection to policing. We then venture in the domain of biometric humanitarianism, examining a large public–private partnership in the same space.

Data Infrastructures: Eurodac From Care to Control

As I write this chapter, a groundbreaking report authored by researcher Antonella Napolitano has just been published by EuroMed Rights, a network of organisations working on human rights protection and democracy promotion in the Euro-Mediterranean region. The report, titled 'Artificial Intelligence: The New Frontier of the EU Border Externalisation Strategy', explores the role of technologies marketed as AI for border externalisation, a widely used strategy to limit migration in EU member states. Discussing the AI act, first voted on 14 June 2023, as a route to regulating artificial intelligence in the EU, the report strikes an important chord on digital ID. Surveillance technology, encompassing 'any digital device, software or system that gathers information on an individuals' activities or communications' (EuroMed Rights, 2023: 9), is becoming an increasingly advanced artefact of border policing. With borders becoming used as the testing grounds for new AI tools, ranging from emotion recognition to speech detection, border management is becoming a growingly attractive market for technology vendors, which supply EU member states with tools to implement migration laws and regulations.

A similar problem has been examined by Bruno Oliveira Martins and Maria Gabrielsen Jumbert, senior researchers at the Peace Research Institute Oslo (PRIO), in terms of the *co-production* of security 'problems' and 'solutions' in the making and commercialisation of EU border technologies. In an article published in the Journal of Ethnic and Migration Studies, Martins and Jumbert (2022) connect migration flows into the EU to increased reliance on technological artefacts in border management. Their study of Unmanned Aerial Vehicles (UAV), most commonly referred to as *drones*, reveals a vision of the migrant as inscribed in a logic of securitisation rather than care and protection. Such an understanding ties in with the arguments of the EuroMed Rights (2023) report; a security solution is co-produced with the logic of migration constructed as an impeding threat, to which novel AI-powered instruments provide a response. Martins (2023) further specifies the point in recent work, where he notes the coupling of knowledge from EU-funded security research and the types of *epistemic control* that the EU itself exerts (Martins, 2023: 436).

The EuroMed Rights (2023) report and the work by Martins and Jumbert (2022) are landmark sources on the role of surveillance technologies in shaping border management, informing its practices and the way people attempting to cross borders, regularly or not, are treated. The enactment of the EU through border management is central in the work of Pelizza (2020), who refers to it with the notion of 'alterity processing': a concept aimed to account for 'the simultaneous enactment of individual "Others" and emergent European orders in the context of migration management' (2020: 262). In the work of Pelizza, alterity processing refers to 'data infrastructures, knowledge practices, and bureaucratic procedures through which populations unknown to European actors are translated into "European-legible" identities' (2020). A conceptual device born in the field of Science and Technology Studies (STS), alterity processing affords investigating the enactment of EU borders through the technologies through which they are policed. This introduces in this book a new type of digital identity systems, aimed not to provide entitlements, but to register and identify people crossing national borders.

Central among such systems is Eurodac, the database collecting fingerprints of asylum seekers for all EU member states. Launched in January 2003, Eurodac is meant to assist EU member states in implementing the Dublin Regulation of 1990; the regulation determines which country is responsible for processing an application for asylum, and states that an application will be processed by the first Dublin country (EU plus Norway, Iceland, Switzerland and Liechtenstein) the asylum seeker comes to. Underpinning Eurodac is the Automated Fingerprint Identification System (AFIS), a digital fingerprint database supplied by the private company Cogent Systems, acquired by 3M in 2010 (3Ms identity management business was subsequently purchased by Gemalto in 2016, prior to the acquisition of Gemalto by Thales in 2019). By storing fingerprints for each asylum seeker, Eurodac prevents people from applying for asylum in more than one member state. It thereby addresses what Thales (n.d.) refers to as 'asylum shopping', meaning the phenomenon for which a person applies for asylum in more than one nation.

A brief history of Eurodac is in order to understand what the database, a data infrastructure based on an architecture for fingerprint storing, effectively accomplishes. When launched, Eurodac had a sharp focus on asylum seekers and on implementing the Dublin regulation. But as noted by Pelizza (2020), in 2005 the European Commission proposed to extend the database to irregular migrants; that is to persons apprehended in the act of an irregular border crossing (Thales, n.d.). The rationale for this shift is again explained by Thales; in the face of the migrant crisis initiated in 1997, crisis talks were engaged by EU member states, leading Eurodac to take on the objective 'not just to record a request for asylum at the first point of entry but also to register those attempting to enter illegally' (Thales, n.d.). If the original goal was that of enforcing an international regulation, the second one, storing fingerprints of people caught in illegal crossing, openly embraces the domain of border policing.

But another shift was to mark the history of Eurodac. It is again Pelizza's (2020) work detailing it; on 20 July 2015, in an almost unnoticed switch, Eurodac was made interoperable with the databases of police forces in all EU member states. A technical shift made through Eurodac's data infrastructure, the switch had strong consequences on surveillance practices; with it, notes Pelizza (2020: 262), police authorities of EU member states could search the database not just to process asylum applications, but also to investigate crimes in their national territory. The changing shape of the Eurodac database, from a technology of care (assisting asylum seekers in the application process) to a technology of control (flagging illegal border crossers and investigating crime), narrates an infrastructural history that is, at the same time, a story of technology repurposing for policing goals.

The infrastructural history of Eurodac illuminates, at least, three aspects of border management that are central to this book. First, works from the Processing Citizenship project (cf. Van Rossem & Pelizza, 2022; Pelizza & Loschi, 2023) clarify how Eurodac is inseparable from the legislative enforcement purpose designed into it. The registration of fingerprints, and the sharing of these among member states, is functional to advancing the Dublin regulation; even before embracing more explicit policing goals, the database used AFIS for defined purposes of legislation enforcement. Thales (n.d.) identifies the limits of the artefact's use; Eurodac compares fingerprints, but 'does NOT store biographic data (name, place of birth, date of birth...)' and 'facial recognition cannot be applied (to it), because the database does NOT include portraits' (capitals in Thales, n.d.). On the one hand, legal enforcement pervades the technology design; on the other, the technology seller's anxiety in clarifying where the system does 'not' reach is meaningful, especially in the light of the accusations received by the EU of being turning Eurodac into a 'mass surveillance tool' (Rankin, 2021).

Such accusations open a second point of relevance. In May 2021, an open letter to the European Parliament, signed by 31 NGOs, expressed vivid concern on plans to extend the Eurodac database well beyond the fingerprint comparison function which it currently performs (Access Now and 30 other signatories, 2021). The letter refers to changes that, agreed on by European Parliament lawmakers and EU interior ministers in 2018, would expand Eurodac to include 'facial images, passport or ID card details, as well as fingerprints. Authorities would also be able to start fingerprinting and photographing migrant children as young as six, compared with the current minimum age limit of 14' (Rankin, 2021). The proposed change is deeply transformative of Eurodac, and seem to explain the anxiety of Thales in highlighting, in capital letters, what the system as it is can 'not' do. At the same time, the striking of the deal in 2018 envisions a database expansion that ventures in much deeper surveillance measures than fingerprinting alone. Extension of the database, argue EU lawmakers, is needed to 'help identify and trace missing children, as part of efforts to protect minors from falling into the hands of human traffickers or smugglers' (Rankin, 2021). At the same time questions of signatory NGOs, including Amnesty International and the European Network Against Racism, illuminate the concern that such an extension, built within a regulative framework named Eurodac Recast, would have on users, building on the interoperability switch that expanded surveillance affordances in 2015 (Access Now and 30 other signatories, 2021).

While at the time of writing, discussion on the biometrics of movement into and around Europe remains open (Hersey, 2023), the history of Eurodac details how surveillant policies, with the outcomes associated to them, are directly inscribed in the artefact's design. The system's evolution can be seen, from the perspective taken by the signatories of the letter to the EU parliament, as an *involution* from care to control; a process marked first by the extension to people apprehended during irregular border crossing, and then, in 2015, by the switch to interoperability with national police databases. If the changes proposed by Eurodac Recast, expanding the system to facial recognition and profiling of migrants from the age of six, were to take effect, the reach of the system into policing would increase substantially, denoting a marked diversion from the 'care' path originally devised. The notion of design-related injustice asserts again how undue surveillance, rather than incidentally interjecting in the artefact's work, constitutes the very flesh of its operational mechanisms (Martins et al., 2022; Martins & Jumbert, 2022).

Finally, technological developments on border policing are in rapid development after the 9 December 2023 agreement on the AI Act, and so is research on the topic. In January 2024 the Border Violence Monitoring Network published 'Decoding Balkandac: Navigating the EU's Biometric Blueprint', a report funded by Privacy International which scrutinises the development of interoperable biometric databases across the Western Balkans. Similarly to what happens with Eurodac in the EU, this region witnesses how 'databases holding information related to people on the move, asylum seekers and beneficiaries of international protection are becoming increasingly interoperable with databases managing information related to crime and security risks'. Focused on countries along the so-called Balkan route, widely used as a path to seek asylum, the report highlights the point by EuroMed Rights (2023) on participation of biometric tools to border externalisation, a strategy of which legislation in the Western Balkans is epitomic (Border Violence Monitoring Network, 2024: 14–15).

A third issue is brought about by the history of Eurodac. Cogent technologies, an US-based software company recently acquired by Thales, held full mandate over AFIS, which it supplies to the EU since 2003. With Thales' acquisition of Cogent, a system deciding on asylum applications – and exerting, as noted here, policing functions on people crossing borders – passed in the hands of the Thales group, which describes itself as 'leader in cybersecurity and data protection' and has a long-term history in building what, under the definition of EuroMed Rights (2023), goes under the term of surveillance technologies. As we enter the terrain of private market interjection into law-enforcing technologies, the question on whose interests are served by digital ID systems, built for the public but manufactured by privates, needs tackling. To answer the question we venture in one of the largest, most contentious humanitarian-tech partnerships of the present times.

WFP-Palantir: Artefacts of Biometric Humanitarianism?

The Eurodac case has shown the results of inscription of surveillant policies into artefacts of border management. But policing is not the only purpose for biometrically identifying people on the move, or subjects made vulnerable by different conditions. The domain of *digital humanitarianism*, which I define here as the assemblage of processes, means and technologies through which the practice of humanitarian work is digitised, is a strong example of how diverse types of digital artefacts are being codified into the humanitarian sector. While the study of digital ID was born, as seen in Chapter 2, as largely state-based, digital humanitarianism illustrates the transition of the same technologies from a national to a supranational domain of action (Martin & Taylor, 2021a; Masiero & Bailur, 2021).

As already noted in Chapter 4, digital humanitarianism is significantly expanding the range of business opportunities for providers of digital ID. Reports on industry fairs such as the Identity Week, which describes itself as 'a conference and exhibition bringing together the brightest minds in the identity sector to promote innovation, new thinking, and more effective identity solutions', are especially prominent on the 'development turn' (Martin, 2021) that digital identity is taking. In a blog post released after Identity Week 2021 in London, Martin and Taylor (2021b) dissected the development logic that ID providers brought to the table. They note how one of the keynotes, representing a large German tech firm, lamented how 'it's a shame that we still have a billion people without a legal identity'. At the same time the digital identity solutions market is forecasted by researchers to grow from US$23.3 billion in 2021 to US$49.5 billion by 2026 (Martin & Taylor, 2021b), and the use of digital identification technology in humanitarian spaces is becoming increasingly propulsive of such a growth (cf. The Engine Room, 2023).

But as a direct consequence of its definition, digital humanitarianism requires partnerships between private technology sellers and supranational organisations, designed for objectives of humanitarian relief. And in the wake of diffusion of such a model, a widely debated partnership of this kind has been that by the World Food Programme (WFP) and Palantir, publicly announced at a press conference on 5 February 2019 (Parker, 2019). Founded in Palo Alto in 2003, Palantir is a data analytics firm with a capitalisation of 34.74 billion as of July 2023, among whose clients are US security agencies and US police departments engaging in data-driven policing (Hvistendahl, 2021, cited in Iliadis & Acker, 2022). Alleged links of Palantir with surveillance, including the digital tracking of undocumented immigrants – including children – for deportation (Iliadis & Acker, 2022: 335), have come under public scrutiny (Amnesty International, 2020; Hvistendahl, 2021); relatedly, the company's revenue stems from a commercial and a government segment, with a total revenue that has increased by 48.79% between the first quarter of 2020 and that of 2021 (Business Quant, n.d.).

Recent work by the Global Data Justice team at Tilburg University (Martin et al., 2023) is illuminating on the move of Palantir into the humanitarian field. With its announcement in 2019, the WFP declared how the five-year partnership was aimed at 'helping WFP use its data to streamline the delivery of food and cash-based assistance in life-saving emergency relief operations around the world' (WFP, 2019a). Martin et al. (2023) note a crucial aspect of the partnership; this involves transgressing the border of the business and security intelligence field, to venture as a large data analytics company into humanitarian terrain. This is what is conceptualised, with Sharon (2021), as a form of *sector transgression* (Taylor et al., 2023). Motivated by the scale and complexity of WFP's operations (WFP, 2019a), the partnership quickly went under public concern, triggered by opacity in the terms of agreement (Responsible Data, 2019) and by the alleged long-established role of Palantir in serving surveillance practices (Amnesty International, 2020; Feuer, 2020).

Against this backdrop, what matters most to a design justice discourse is one of the least discussed aspects that the partnership entails. Animated by Costanza-Chock's (2020) focus on the artefact, we are especially concerned with the core of the technology on which the partnership is built, and the nature of the practices that such a technology entails. WFP's statement (2019a) is clear on this aspect: 'WFP will use Palantir's Foundry software platform to bring together data sources from across the organization, enabling staff at headquarters and in field locations to access and analyse programmatic and operational data in a secure, unified environment'. Foundry is one of the three platforms that, with Gotham and Apollo, constitute Palantir's ecosystem; Gotham is described by Palantir (n.d.) as 'a commercially-available, AI-ready operating system that improves and accelerates decisions for operators across roles and all domains'. Built on top of Gotham, Foundry is described as 'a platform for managing business complexity', which 'provides flexibility for both technical and non-technical users to integrate, manage and interact with their data – and create lasting business value from those inputs' (Palantir, n.d.).

Now, it is important to understand what a platform for data integration – largely used, beyond the business domain, for intelligence purposes – can bring to the WFP. Perhaps the most informative statement on the topic came from Enrica Porcari, Chief Information Officer at the WFP, just two days after the partnership's announcement. The statement answered the concern raised from multiple public voices on data privacy. In the face of questions on how data from over 90 billion beneficiaries of the WFP would be handled, Porcari asserted that rules of engagement (with all partners of WFP) included 'no access to WFP data that provides beneficiary information', and that 'any data analytics partner we work with understands clearly that this is our starting point and we are unwilling to compromise' (WFP, 2019b). She underscored, at the same time, the rationale for the partnership, lying in how 'our mission is to serve the needs of the 821 million hungry people in the world, and we have been tireless in our search for partnerships with industry leaders in the private sector who can enhance our reach and improve our delivery of humanitarian assistance' (WFP, 2019b).

But how does the WFP use the instruments of data analytics that Palantir provided them with? The WFP statements (2019a, 2019b) do not offer detail on this, a point that an open letter signed by 69 among activists, academics and NGOs has noted (Responsible Data, 2019). At the same time, WFP (2019a) contained an important piece of information; the partnership was built upon a pilot project which aimed at supporting Optimus, the WFP's tool for supply chain optimisation. The recipient of the Franz Edelman Award for analytics and optimisation in 2021, Optimus is described on the WFP website as a decision support system that 'helps identify the most efficient and cost-effective way to reach the people we serve'; it notes how 'building on data from dozens of sources, mathematical models provide key insights into food basket design, sourcing strategies, and delivery networks for any WFP operation'. Having saved WFP more than US$30 million in the operations where it is being used (WFP, 2019a), Optimus is in line with the data integration functionalities of Palantir's Foundry, and affords the ability to transpose Foundry's business data analytics on decision-making to achieve programmatic goals (for example, on nutritional values and costs) within operational constraints, for example on funding levels and supply chain lead times (WFP, 2019a).

It is here that a design justice problem emerges. Relying on Taylor et al.'s (2023) lexicon, Palantir's *sector transgression* means that the business logic of Palantir, which is well clarified in the company's documentation of Foundry (Palantir, n.d.), ends up embedded in a software which should be centred on humanitarian principles, such as those of causing no harm (Manby, 2021) and that 'humanitarian actors must not take sides'. But such principles, continue Martin et al. (2023: 1363), seem at odds with Palantir's CEO statement that 'the core mission of our company always was to make the West, especially America, the strongest in the world, the strongest it's ever been, for the sake of global peace and prosperity' (quoted in Feuer, 2020). More at large, they seem at odds with Palantir's alleged history of involvement in predictive policing (Sherman, 2020; Hvistendahl, 2021). A dynamic in which, note Martin et al. (2023: 1363), WFP's operational data will provide Palantir with core insights on global food (in)security, raising the questions on how such insights will be used.

Palantir's history is especially problematic in the light of this point. In the report 'Failing to Do Right: The Urgent Need for Palantir to Respect Human Rights', Amnesty International (2020) reports on the centrality of two Palantir artefacts, the Integrated Case Management System (ICM) and FALCON analytical platforms, within human right violating acts seeing asylum seekers and migrants divided from family members, subjected to workplace raids, detained and deported by the Immigration and Customs Enforcement (ICE) of the United States. Using ICM, notes Amnesty International (2020: 4), ICE can access a wide range of personal data from US agencies and law enforcers; the platform acts in combination with FALCON, which combines a 'search and analysis' function with a tool to 'link information from tips to existing databases, and create profiles based on those tips' (2020: 4–5). The combination of ICM and FALCON, continues the report, has been utilised to harm migrants and asylum seekers. In 2017, ICE used ICM technology to arrest over 400 people in a detention-leading operation targeting parents and caregivers of unaccompanied children. FALCON technology has been equally used to plan mass workplace raids, with a raid in Mississippi that, in 2019, led to the arrest of nearly 700 people and the separation of children from parents and caregivers (Amnesty International, 2020: 6). Thinking of Palantir as an entity in charge of an information system for global food distribution is especially concerning in the light of such reports, and of the direct evidence of data-induced harm they detail.

With the design justice light that guides this chapter, two arguments can be drawn. First, Palantir's history of sector transgressions remarks how artefacts, even in the allegedly apolitical humanitarian space, embody a system of interests that is reflected in the way the technology plays out. Under the rubric of savings, cost efficiency and better assistance for the underserved WFP beneficiaries, Palantir is provided with core insights on food (in)security, insights whose utilisation – from a data analytics company with a strong business focus – leaves many questions open. Secondly, in a discourse centred on companies whose large capitalization inspires a vendor focus, the perspective of the recipient remains important, and one that is resoundingly absent from the discourse on biometric vendors today (Martin & Taylor, 2021b). As aidwashing – defined, with Martin (2023), as laundering a company's reputation with involvement in aid – enters the discourse, what is reflected here is the experience of recipients profiled, policed and deported, whose voices risk being unheard in the discourse on biometric markets and their rapidly increasing capitalisation.

The Dark Matter of Digital ID

Based on the design justice principles arrived at in the AMC, Costanza-Chock (2020) noted how injustice, and the harm stemming from it, can be inscribed directly in the way artefacts are designed. This point has contributed to a substantial evolution in the concept of design-related injustice from Masiero and Das (2019). In noting the discrepancy between PDS users' need for access and an artefact associated to erroneous exclusions, we had thought of the issue as a *design-reality gap* (Heeks, 2002) in which designers did not share the same priorities as users. Yet, with her work on injustice scripted into artefacts, Costanza-Chock (2020) inspired a rethinking of the concept; rather than just a gap in priorities, injustice can be predicated directly on how the artefact works. If it is so, as noted above, we are not looking at a dark 'side' of technology, but at injustice embedded in the very way in which the artefact is built and operates.

This has led me to argue, on the pages of the Electronic Journal of Information Systems in Developing Countries (Masiero, 2023), that a concept of *dark matter* should be juxtaposed to the widely popular notion of a 'dark side' of IT. On the one hand, IS literature is widely diffused on the 'dark side'; topics including techno-stress, IT interruptions, IT misuse or overuse and information security breaches (Tarafdar et al., 2015) point to 'side' effects of IT, which can be corrected with incremental technical improvement. With the concept of *dark matter* I intend, instead, a situation in which injustice lies at the very heart of the technology; rather than peripheral, it is scripted into the technical features of the artefacts, which is therefore inseparable from the harmful effects it generates. The airport scanners discussed by Costanza-Chock (2020) are illustrative of this; injustice towards people whose body does not conform to gender stereotypes is not a 'side' effect of the scanner, but the very way it is programed to work. In the revised concept of design-related injustice, the problem is not anymore just a gap in priorities, but the injustice inscribed in the technology's features.

All three cases discussed in this chapter can be read in the light of a *dark matter* of digital ID. In the Aadhaar-based PDS, technology is constructed with the purpose of monitoring users, so that persons that are not recognised as entitled beneficiaries cannot access rations. Doing so, the system prioritises a type of error – inclusion error – that does not, on the other hand, address the plea of the 'excluded needy' (Swaminathan, 2002) that unsuccessfully seek access to a system they are entitled to. The injustice stemming from this is no 'side' effect; as we have seen in Chapter 4, it is the product of two decades of Indian food policy history, in which leakages from the PDS have been seen as the main problem impeding a successful food security scheme. The result is a technology scripted to prioritise inclusion errors, but that does not offer any mitigation mechanisms for the unjustly excluded; while our 2014 fieldwork showed examples of ration dealers recording transactions on a copybook in case of machine failure (Masiero & Prakash, 2015), we have not seen this option with ABBA in 2018.

In the case of Eurodac, an involution from care to control is inscribed in multiple phases of the system's history. When launched in January 2003, the system was squarely centred on asylum seekers; but already in 2005, it became extended to irregular migrants, augmenting a database of asylum seekers with records of people apprehended in the act of an irregular border crossing. Such policing functions became more pronounced when, in 2015, Eurodac became interoperable with databases of police authorities in all member states. The plan of Eurodac Recast, which stretches into identification of minors from the age of six and into practices as invasive as face recognition, goes even deeper into surveillance. With this additional turning point, the artefact's history speaks clearly of a technology that, born with a logic of care (correctly storing asylum seeker data in processing their applications), evolved into a more and more pronounced logic of control. Surveillance features, and the harm derived from them, are again not a 'side' of Eurodac; they are directly embedded in the infrastructural history narrating how the artefact has evolved.

In the case of Optimus, the WFP supply chain management tool now supported by Palantir, the problem is written in the humanitarian-tech partnership that the WFP and Palantir have created. On the one side is the WFP, with its humanitarian logic and the need to manage an extremely complex supply chain, catering to over 90 million users worldwide. On the other is Palantir; a tech giant that, with its enormous capacity of big data analytics, has propelled acts of surveillance (Amnesty International, 2020). The choice of Palantir as a partner for the WFP has been criticised on ethical and data protection grounds (cf. Responsible Data, 2019). From a different, design-related justice angle, the issue lies in the infusion of Palantir's surveillance logic into an artefact designed to protect vulnerable people with food aid. Once again the issue is not with a 'side' but with the heart of how the system works; the open question, on how WFP data will provide Palantir with global insights on food insecurity, raises more concerns in the light of this logic.

Two points should be made on how a *dark matter* notion can help us navigate the issues of design-related injustice. First, other notions in the literature have pointed to the conflation of care and control in artefacts, and have applied the same notions to technologies of identification that, deployed for humanitarian goals, are plied to a surveillant function. As noted in Murakami Wood and Firmino (2009: 299), the view of the spectrum of citizenship as comprised within the poles of the 'caring, inclusive, rights-based' society and the repressive, control society has pervaded surveillance studies (cf. Lyon, 2007). Gianluca Iazzolino, Lecturer in Digital Development at the University of Manchester, draws on Fassin (2005) in unpacking the concept of *infrastructures of compassionate repression* to capture the tensions created by the Biometric Identification Management System (BIMS), a technology for refugee registration implemented by the UNHCR in Kenya. Drawing on field-work in Kakuma, one of the largest refugee camps close to the Somali-Kenyan border, Iazzolino notes the propositive enthusiasm of the UNHCR when BIMS was introduced. Such enthusiasm was met, at the same time, with anxiety and widespread concern by Somali refugees in the camp. Concerns went to the extent that, Iazzolino (2021: 111–112) reports, refugees proposed to negotiate the acceptance of BIMS in exchange for larger food packages, to somehow compensate for the risks that the new biometric registration system was seen to involve.

Iazzolino's (2021) ethnography, conducted with Somali refugees in Kakuma, illuminates the roots of the widespread concern felt and expressed towards BIMS. The system emerged in the context of a long-term, tense relation across Kenyan and Somalian borders; as noted in Chapter 4, suspicion and violence affected the ethnic Somali population in Kenya, at least since the country's independence from Britain in 1963 (Weitzberg, 2017; Iazzolino, 2021: 112). The objects of systematic profiling and policing, Somali refugees interviewed by Iazzolino expressed both surveillance and practical anxieties on BIMS. Surveillance because the system, associating biometrics to demographic data, was seen as a route for Kenyan authorities to enact violent measures, culminating into deportation. Practical because a biometric system aimed at determining refugee household entitlements has a policing logic scripted in it, as well laid out by a UNHCR official that Iazzolino (2021) interviewed:

> It is a matter of fairness, because not everybody is entitled to aid. Those who are claiming aid without being refugees, or using the card of someone else are robbing other refugees. Iazzolino (2021: 120)

The official refers to Kenyan people of Somali origin who passed themselves as refugees, to access vital supplies of food, shelter and education (as noted when discussing double registration in Chapter 4). From the official's logic, needy people accessing camp services would be 'robbing' other refugees; the policing side of compassionate repression, residing in the use of biometric registration to ensure entitlement, would arguably find justification in this point. But at the same time, the point obscures the plea of people who, deprived of the most essential assets, sought access to refugee camps in the look for basic supplies, extremely hard to reach as a non-refugee (Haki na Sheria, 2021). Illuminating the indivisibility of care and control in the making of biometric artefacts, Iazzolino's infrastructures of compassionate repression are a guiding light in our analysis of injustice.

Secondly, the notion of compassionate repression is an important device in navigating the issues of border policing, apprehension in border crossing and treatment of people on the move that this chapter has analysed. The EuroMed Rights (2023) report is clear on this aspect; EU-funded projects are in the process of creating experiments with advanced surveillance and predictive tools, including instruments with the designed goal to 'forecast future mobility' (EuroMed Rights, 2023: 16). Such projects, the report continues, include initiatives to help humanitarian and migration authorities plan and allocate their resources; one example is EUMigraTool (EMT), which according to its website 'will provide predictions of the number of migrants coming to a specific European country, "analysis on drivers, patterns and choices of migration, as well as public sentiment towards migration"', and 'the identification of risks of tensions between migrants and EU citizens'. The tool, the report notes, has been opposed by civil society and academics (Access Now and 18 other signatories, 2022). Its description is revealing of a predictive policing ontology, endangering basic data protection rights and where *compassionate repression* shifts strongly towards the policing side of the term.

And at the same time, repression is only one part of the story that this chapter has told. Let us not forget the history of the pre-Aadhaar, PDS-enhancing machine designed under the Harish Gowda administration; short-lived in the face of the incoming ABBA, the system sought to develop a stronger PDS, fixing the impediments that stemmed from private outsourcing of the ration card process in 2009. The system was not made to dismantle the PDS; on the contrary, it was devised to reach the best outcomes under subsidisation, as opposed to the cash transfer logic implanted in Aadhaar since the envisioning of JAM (Government of India, 2015). While the *dark matter* notion points at injustice designed into technology, justice can be designed in it as well, and it is with this idea that we move to the forms of *fair ID* that, having taken stock of injustice, the next section explores.

Summary

Chapter 6 has brought us to the end of the journey across injustice that Section 2 consists of. We have explored legal injustice, which we defined as *the conditionality of legal rights and entitlements to digital identification and authentication*. We have then moved into the hidden layer of informational injustice, meaning *injustice perpetrated through the obscuration of information on use of data from digital identification*. And with this chapter, we have explored design-related injustice, noting what happens when injustice, rather than peripheral or incidental, is directly scripted into digital ID technology. The three types of injustice interact with each other, and in the next section we will interrogate how people organise in responding to them.

We have gone through multiple injustices at this point. And yet, with this chapter we have seen that not only injustice, but even virtuous goals – virtuous, in the first place, for technology users – can be inscribed in the systems we are looking at. To zoom on such virtuous goals, we move to the exploration of resistance efforts to the harm caused by digital ID.

PART 3
RESISTANCE

7

ON ID, SOLIDARITY AND RESISTANCE

As this book has shown so far, the orthodoxy connecting digital ID to development is countered by multiple reports of how digital identity affects people's lives. With the importance of ID transcending into the domains of social protection and humanitarian assistance, reports increasingly underscore the ways in which different forms of ID can result into more or less direct harm on people. In this space, increasing attention is given to the use of digital ID beyond social protection and services, in spheres including humanitarian provisions and the work of supranational organisations.

Different aspects of such harm are captured across public reports. As noted in Chapter 4, in June 2022 a team led by Katelyn Cioffi, Victoria Adelmant and Christian Van Veen, from the Centre for Human Rights and Global Justice at New York University, published 'Paving a Digital Road to Hell: A Primer on the Role of the World Bank and Global Networks in Promoting Digital ID', a report that identifies World Bank sponsored digital identity programmes as the source of multiple violations in the human rights space. The report points to a new 'development consensus' centred on digital ID: a consensus that the ID for Development initiative of the World Bank, as part of a broader network of interests and powers, is uniquely positioned to advance through programme funding. These systems, argues the report, provide individuals with a transactional or 'economic' identity. This can be carried around, and enable access to programmes and services while at the same time constructing a real digital economy. This model is associated to one particular type of schemes, where ID is centralised and needed for essential transactions to take place.

The report, however, points to evidence related to large-scale human right violations connected to this form of digital ID. Reviewing digital ID systems from across countries, it points to exasperation of multiple forms of exclusion and discrimination in public services, under the guise of what Chapter 4 has termed *legal injustice*. It also notes the new forms of harm linked to digital identity systems, affirming the highly significant role played by these in relation to surveillance capitalism. As a result, the report is a critique to the new economic paradigm advanced by centralised digital ID systems; it calls actors in the human rights space to ask what can be done to 'ensure that digital ID systems enhance, rather than jeopardize, the enjoyment of human rights' (Centre for Human Rights and Global Justice, 2022: 5). The research published in the report resulted into an open letter signed by NGOs, academics and members of the civil society, demanding the World Bank and its donors to cease activities that promote harmful models of digital ID, while also enforcing greater transparency around activities of the World Bank in the digital ID domain (Access Now and 45 other signatories, 2022).

But reports study digital ID beyond its effects at the state level. In July 2023, the non-governmental organisation The Engine Room released a report called 'Biometrics in the Humanitarian Sector', whose focus is on biometric systems deployed for recipients of humanitarian assistance to access vital forms of aid and support. The evidence presented in the report leads to one key take-home; while evidence of benefits of biometric humanitarianism is weak and scattered, new pathways to harm have emerged from the same technologies. Through cases of data collection practices associated to refugee repatriation (for instance, in the case of Rohingya refugees in Bangladesh), as well as new predicaments resulting from the denial of an ID card (as it is for double registered persons in Kenya), the report argues that the sector is 'slowly deepening' awareness of harm associated to biometric systems, but is responding to such risks in mostly fragmented ways (Haki na Sheria, 2021; Privacy International, 2021). Such fragmentation, concludes the Engine Room's (2023) report, still needs serious addressing by international juridical and humanitarian entities.

The report also notes how biometric humanitarianism blends with border control, intersecting with hostile environments that magnify the problems lived by digitally profiled people (Latonero et al., 2019; EuroMed Rights, 2023; Border Violence Monitoring Network, 2024). The role of digital identity in border policing, studied in Chapter 6, is central to 'Europe's Techno-Borders', released in July 2023 by the UK charity Statewatch. Focusing on the digitalisation of borders within Europe, the report looks at its effects on people trying to enter the EU, or to stay in its territory; it finds digital technologies underscoring 'invasions of privacy, brutal violations of human rights, and make the border "mobile", for example through the increased use of mobile biometric identification technologies' (Statewatch, 2023: 5). The analysis extends to EU public funding of border technologies, which increased by 94% across two budgetary periods (Statewatch, 2023) and underpins the multiple human rights violations that the report details.

All these instances of harm motivate the focus on injustice perpetrated through digital ID that the book has taken so far. An important aspect of the reports mentioned here is the complementarity of their objects; looking respectively at state-centred ID systems, biometric humanitarianism and border management, reports integrate each other in painting a picture of oppression rather than liberation through digital ID. These are not only reports on the ID systems themselves: all of them study the impact of digital ID on people's lives in situations of vulnerability, ranging from the need to access social protection to that of availing humanitarian supplies or flee war and violence. Transversal across reports is also the finding that benefits of digital ID are weakly documented, unlike harms for which there are, instead, strong bases of evidence across countries. In the light of this book, all these works point to the question on 'what happens to the user' as a result of digital identity, enriching the picture of injustice that the previous section has painted.

And at the same time, there is a part of the story that complements injustice. With the sustained global diffusion of digital ID, research on it reveals another overarching aspect; faced with the realities of harm discussed in this book, people affected by such harms aggregate with each other, finding ways to voice their predicament and imagine alternatives to it. Resistance to ID-induced harms results into practices of solidarity making, which contrast artefacts of oppression (such as the *dark matter* systems reviewed in Chapter 6) with different legal, technical and political routes to fairness. It is on such solidarity practices, and on their role towards imagining views of 'fair ID' that contrast injustice, that we focus in this chapter and in this final section.

The lens of *data activism* is crucial in illuminating these solidarity practices. It is from this lens that the chapter begins, noting the ability of a data activist perspective in illustrating community interventions occurring in the digital ID space. Stories associated to India's Right to Food campaign, the work of Kenya's human right organisations towards deregistration of double registered people, and the active commitment to ID fairness collectively built through the #WhyID campaign, enact a data activist lens in the narration of day-to-day practices of resistance to unfair ID. I argue that it is through these practices, and through their constructive potential, that new forms of 'fair ID' can be imagined.

Before moving on, a crucial point must be made. The three cases narrated here have been selected with an illustrative purpose, identifying in them the traits of what below is termed proactive and reactive data activism (Milan & Van der Velden, 2016). The choice of these cases does not come with intentions of granting prominence over other realities; it does, on the contrary, illuminate traits of data activism that are transversal in movements responding to unfair ID. In studying such traits, the light is also cast on the multiple inter-organisational partnerships that such realities undertake.

Data Activism and ID Resistance

From the opening of this book, a central point arose on the role played by data in the making and substantiation of digital ID. Defining digital identity as the conversion of people into machine-readable data brings a direct bearing on the architecture of digital ID systems; it is through (demographic and biometric) data that people are registered into ID databases, that they authenticate at points of access, and that they are authorised, or not, to access given services. The 'data flesh' of digital identity systems has been key in studying their architecture; it has inspired the data justice lens used in this book, through which we have met three different, but unescapably data-based forms of injustice.

But data, it is time to note, also pave the way for responding to injustices. And if a data justice lens has helped us make sense of unfair ID and of the different shapes that it takes, it is time that a sister lens, knowable with the term of *data activism*, also enters the conversation. On the one hand, data injustice helps us illuminate unfairness in how people are visualised, represented and treated through their production of digital data (Taylor, 2017). But on the other, data activism works in a specular way; the term came to indicate a family of resistant approaches, where the very technologies used to affirm 'power over' people are plied to mobilisation for a range of social objectives (Treré, 2018; Couldry & Mejias, 2019). If data injustice gives us a lens to understand data-induced unfairness on the lives of people, data activism provides the conceptual means to understand how people can leverage data to acquire, or regain, agency and control over central aspects of their datafied lives.

The concept of data activism has had multiple characterisations. The works of Renzi and Langlois (2015) and Schrock (2016), cited in Milan and Van der Velden (2016), offer core building blocks of its definitions; in situating the concept in the realm of collective action, Renzi and Langlois (2015) define data activism as 'new modes of being and acting together through a direct engagement with data and the means of its mobilisation'. The duality of data in their essence and the means to which they are put to action is also present in Schrock (2016), who characterises civic hackers as individuals who 'operate through a range of data-driven political modes [...] to bring about systemic change [...] [and] participate in civic data politics' (2016: 591, cited in Milan & Van der Velden, 2016: 60). In both definitions of data activism, data emerge both as a means and an end to liberating missions; as a means because it is through data that, for instance, civic hackers contribute to liberating communities from oppressive behaviours (Fussy, 2021), and as an end, because the byproduct of activist practices can coincide with fairer forms of access to data for subjects.

But when applied to resistance in digital ID, the notion of data activism needs a shift away from the civically shaped, hacking-centred definition that much of the early literature clusters around. This book finds that shift in Milan and Van der Velden (2016); in their groundbreaking piece 'The Alternative Epistemologies of

Data Activism', they define data activism as indicating 'the range of sociotechnical practices that interrogate the fundamental paradigm shift brought about by data-fication'. Such sociotechnical practices reflect the new forms of societal engagement that the datafied society has brought; it is their sociotechnical nature that makes data an ensemble in which people partake, both by acting *on* data and *through* data themselves. In this way, the vision introduced by Milan and Van der Velden (2016) extends the concept of data activism from particular forms of hacking-centred volunteerism to a discourse in which people, affected by different forms of digi-tally induced harm, aggregate in order to collectively respond to it.

The extended notion provided by Milan and Van der Velden (2016) is articulated on a continuum of data activist practices. The way the continuum is structured is especially powerful in understanding the practices that the concept entails; *proactive* data activism, they note, refers to different ways in which people perform affir-mative engagement with data. In the datafied society, multiple forms of such pro-active, data advocacy-centred engagement can be observed. In a recent instantiation, the multilingual volume 'COVID-19 from the Margins: Pandemic Invisibilities, Policies and Resistance in the Datafied Society' (Milan et al., 2021) offered multiple examples of how data journalism offered ways to challenge the pandemic policies of selected governments. In one of the chapters contained in the book's fifth section, programmatically titled 'Pandemic Solidarity and Resistance from Below', Fussy (2021) details the case of data volunteerism in Brazil, where governmental manipulations of COVID-19 data visualisations were actively coun-tered by grassroot activism, seeking to provide open access to real time and truthful data on the evolution of the pandemic.

At the other end of the continuum is *reactive* data activism, which Milan and Van der Velden (2016) define as 'tactics of resistance to massive data collection'. In this notion, the focus is not anymore on what can be done 'with' the data; it is instead, in a specular way, on reaction to data practices seen as unfair, practices that often coincide with the indiscriminate data collection required for enabling access to a plethora of services. This notion is reminiscent of Kitchin (2014), who notes that a central feature of big data is, beyond size, the ability of 'being generated continuously, seeking to be exhaustive and fine-grained in scope, and flexible and scalable in its production' (2014: 2, cited in Milan & Van der Velden, 2016: 60). In reactive forms of data activism, the continue generation of data about subjects is openly recognised as a source of unfairness. The indiscriminate production of data on people enables the very forms of design injustice that Chapter 6 has illuminated, for instance by cross-checking data of asylum seekers with policing authorities (Pelizza, 2020) or enabling extreme forms of AI-based border management (EuroMed Rights, 2023). It is the unfairness streaming from indiscrimi-nate data collection that animates reactive data activism, and that makes it an integral part of the data-centred practices of resistance that we see today.

The expansion conducted by Milan and Van der Velden (2016) on the notion of data activism is at the basis of the use of this conceptual device in the exploration of resistance to digital ID. On the one hand, a substantial part of ID resistance qualifies as action taking place *on* the data artefacts that lie at the origins of the forms of injustice that this book has studied so far. Action on digital identity systems, and on the ways they enable the authorisation-authentication nexus at the basis of digital ID, forms a central route in terms of how ID unfairness is countered; an example lies in the work of human rights organisations that, in Kenya, take action towards deregistration of double registered people from the refugee database (Haki na Sheria, 2021; Mutung'u, 2021). In this case it is the direct action on the artefact – the database – that enables people's ability to receive an ID card, in virtue of not anymore having a duplicate record in the refugee registration system. Similarly, one of India's largest food activism movements – the Right to Food campaign – disputes the use of Aadhaar in food security programmes in order to break the legal injustice circuit that Chapter 4 has illustrated, and that, as it will be shown below, extends well beyond the PDS, down to schemes that provide midday meals and Integrated Child Development Services (ICDS).

On the other hand, proactive forms of data activism feature prominently in the digital ID space. These forms move beyond data advocacy: for instance, India's Right to Food campaign sustains certain types of cash transfers – based especially on maternity entitlements – as a route to advance a food secure society. In open problematisation of the view that cash transfers are inherently good or suboptimal for recipients, the campaign relates this measure to the situation of targeted users; there are data-enabled cash transfers that can result in fair outcomes (Privacy International, 2018), which should be separated by the indiscriminate adoption of Aadhaar in social protection schemes (cf. Drèze & Khera, 2015; Khera, 2017, 2019; Muralidharan et al., 2020). Similarly, the fundamental question on 'Why ID?' does not only enquire 'why' essential forms of sustenance should be conditional to ID in the first place, but also notes how misdeeds conducted under the aegis of 'Big ID' can be countered by grass-root work (Privacy International, 2019a, 2019b; Access Now, 2021).

As noted by Milan and Van der Velden (2016), the extended notion of data activism is a heuristic tool, to be adopted as a route to conceptualising reactive and proactive ways to engage with data injustices. In what follows I operate such a tool, using the extended notion of data activism to work through the practices of three protagonists of the fight to unfair ID on a global scale.

India's Right to Food Campaign: (Fair) ID as Gateway to Food

While conducting my PhD fieldwork in Kerala in 2011/2012, the work of India's Right to Food campaign was repeatedly signalled to me as a paradigm of community activism for food security. This was shortly before a paper by Dipa Sinha – an activist within the Right to Food campaign, and currently an Assistant Professor at Ambedkar University, Delhi – captured my attention in the *Economic and Political Weekly*, a popular social science journal published by the non-governmental network Sameeksha trust. Titled 'Cash for Food: A Misplaced Idea', Sinha's (2015) paper took issue with the idea that direct benefit transfers (DBT) could replace the PDS. Her piece brought to light the issues presented by DBT versus food, especially the inability of DBT to identify ('target') the poor and the risk of subjecting below-poverty-line (BPL) recipients to price fluctuations. Published during my research on the PDS in Karnataka, Sinha's paper spurred my interest for the Right to Food campaign's work, which squared with my focus on understanding the impact of different digital solutions on food security systems.

The campaign's foundation statement (Right to Food India, n.d.) illuminates its central purposes. As noted in the statement, the Right to Food campaign is 'an informal network of organisations and individuals committed to the real-isation of the right to food in India'. Recognising such commitment as the network's constitutive basis, the statement proceeds to outline the tenets of the campaign:

> We consider that everyone has a fundamental right to be free from hunger and undernutrition. Realising this right requires not only equitable and sustainable food systems, but also entitlements relating to livelihood security such as the right to work, land reform and social security. We consider that the primary responsibility for guaranteeing these entitlements rests with the state. Lack of financial resources cannot be accepted as an excuse for abdicating this responsibility. In the present context, where people's basic needs are not a political priority, state intervention itself depends on effective popular organisation. We are committed to fostering this process through all democratic means. (Right to Food India, n.d.)

The campaign (Right to Food India, n.d.) began in 2001, stemming from a public interest litigation in India's Supreme Court. A writ petition was submitted by the Supreme Court by People's Union for Civil Liberties, Rajasthan (PUCL vs Union of India and Others, Writ Petition [Civil] 196 of 2001); its central demand was that the country's large food stocks were to be 'used without delay to protect people from hunger and starvation'. The petition spurred a prolonged public interest litigation, involving Supreme Court hearings and several interim orders (Right to Food India, n.d.). The need for supporting the legal process inspired a collective effort to organise for the right to food, leading to the creation and sustenance of the campaign.

With its over 20 years of activity, the Right to Food campaign has engaged multiple cases. As per the campaign's website (Right to Food India, n.d.), examples include 'public hearings, rallies, dharnas, padyatras, conventions, action-oriented research, media advocacy, and lobbying of Members of Parliament'.[1] Demands include universal midday meals in primary schools, universalisation of ICDS for children under the age of six, effective implementation of all nutrition-related schemes and an overall revival of the PDS across the country (Right to Food India, n.d.). Defending the right to food through the multiple schemes that support it, the campaign is key to the point that not just the PDS, but a larger set of programmes – to which midday meals and ICDS are central – contribute to the articulate armoury of India's food security system.

The campaign's founding statement does not mention data or digital ID. It is a campaign centred on food as a fundamental human right; but exactly in this guise, the campaign became a key entity to engage the digital mediation of India's food security programmes, through digital ID and other means. Central to this engagement was the mediating role of Aadhaar-enabled authentication, electronic weighing scales in ration shops and other technologies for access to food-related entitlements, which informed the campaign's action in both proactive and reactive ways. In what follows I shed light on some key works from the campaign, centring on different routes undertaken to defend the fundamental right to food.

Reactive Focus: Opposing Aadhaar-Based Exclusions

One way to learn about the work of the Right to Food campaign is through the open access primers published on its website (Right to Food India, n.d.), focused on different aspects of food-related programmes and legislation. One of these primers concerns the National Food Security Act (NFSA), passed in 2013 as a route to legal recognition of food as a fundamental right. The petition from which the campaign stemmed argued that the right to food derives from the fundamental 'right to life', enshrined in Article 21 of the Indian Constitution. The NFSA was an essential step in translating the right to food into law; it defines three basic sets of entitlements to (a) subsidised food grains under the PDS, (b) nutritious food for children, pregnant and lactating women and (c) maternity entitlements for pregnant women. Specific entitlements, centred on particular groups of users, are devised for all three categories.

---

[1]*Dharna* refers to a form of protest centred on sitting and not leaving a place in demand for justice. *Padyatra* refers to a journey undertaken on foot, which mobilises people together for collective societal purposes.

Especially the points on 'nutritious food' and 'maternity entitlements' go beyond the PDS, a core food security programme on which this book has focused so far. On nutritious food, entitlements state 'free supplementary nutrition (cooked meals or "take-home rations") for children below the age of 6 years at the local Anganwadi', where *anganwadi* is a type of childcare centre launched as part of the ICDS (exclusive breastfeeding is promoted by the NFSA for children below six months). They also refer to a 'hot, cooked, nutritious midday meal every day for school children up to Class 8 or within the age group of 6–14 years (whichever applicable)', and to 'one free, nutritious meal every day from the local Anganwadi for women during pregnancy and six months after childbirth, in the form of take-home rations' (Right to Food India, 2016a: 9–10). Maternity entitlements consist, instead, of cash; they are 'cash grants of at least Rs 6,000 per child (in instalments), for every woman not already benefiting from maternity entitlements under other laws or through regular government employment' (Right to Food India, 2016a: 10). These entitlements show that, notwithstanding the importance of the PDS, a set of supplemental integrated services is fundamental for food security in India, targeting categories in special need for sustained good nutrition.

Many actions have been taken by the campaign in defence of the country's food security schemes. But information, the Right to Food India (2016a) NFSA primer notes, is crucial to many of these actions; the NFSA expanded entitlements under multiple schemes, requiring the PDS to cover at least 75% of the rural population and 50% of the urban population at the all-India level (Right to Food India, 2016a: 11). It is crucial that households are informed of the entitlements that apply to them; and technology, the primer continues, is central to spreading such information. Every ration shop, as discussed in Chapter 5, must have a complete information board, indicating user categories and the respective entitlements. In addition, electronic weighing machines in ration shops can support proper information, for example by displaying the exact quantity of food grain disbursed in each transaction. Section 12 of the NFSA instructs central and state governments to undertake necessary PDS reforms; these include doorstep delivery of food grains to PDS outlets, full transparency of records, preference of public institutions in ration shop licencing, and 'application of information and communication technology tools including end-to-end computerisation in order to ensure transparent recording of transactions at all levels' (Right to Food India, 2016a: 23). Such a phrasing invites the idea that technology can support proactive PDS reforms designed to benefit the user.

It is in this primer that the Right to Food campaign makes one of the clearest statements on how technology can help, rather than hinder, the achievement of the right to food for all. The primer (Right to Food India, 2016a: 24) notes that 'computerisation of ration cards helps to impart transparency and prevent corruption'; this supports the point made by Karnataka's former Secretary for Food, Civil Supplies and Consumer Affairs, who wanted to strengthen the PDS through digitised records for each beneficiary household (as seen in Chapter 6). The second point is on electronic weighing machines in ration shops; these contribute to the spreading of information, as the machine offers an electronically accurate depiction of what is being sold to the user. While manipulations are possible, for example by muting the machine speakers that in Karnataka announced the quantity of goods being sold (Masiero & Prakash, 2015); the inner logic of the artefact is that of providing important information to the user at the moment of the ration shop transaction.

The informative role of digital ration card records and electronic weighing machines played a central part in the incorporation of digital technologies in PDS reform. Shortly after the publication of Sinha's (2015) paper in the *Economic and Political Weekly*, I published in the same journal a short piece, titled 'Will the JAM Trinity dismantle the PDS?', in which I questioned the idea that Aadhaar and related technologies are necessarily to be used to promote a shift to cash transfers. The paper used examples from the states of Kerala and Karnataka to show an opposite point; namely, that technology can result in actions that strengthen food security systems, such as the pre-Aadhaar technology routes that both states undertook in relation to the PDS (Masiero, 2015b). Given the proven importance of reforms in improving offtake and user experience in the PDS at the state level (Khera, 2011b; Drèze & Khera, 2015), it is essential to study the role that datafying technologies can have in it. As it happened in relation to Aadhaar, datafying technologies can participate in the exclusion of entitled users; but as shown in the stories of the PDS in Karnataka and Kerala, they do not have to, as similarly designed technologies can participate in food policies that support user access to entitlements.

But it is with respect to Aadhaar, and especially its incorporation in food security systems, that the campaign shows strong reservations. The NFSA primer (Right to Food India, 2016a) already notes the issue, pointing at the problematic incorporation of Aadhaar in the ration shop's customer interface:

> However, some reforms introduced under Section 12 [of the NFSA], such as the imposition of Aadhaar-based biometric identification over the internet using "Point of Sale" (PoS) machines, are proving very disruptive. This system requires many fragile technologies to work at the same time: the PoS machine, the internet, the fingerprint recognition process, and the mobile network. In addition, the data groundwork (including Aadhaar seeding) must have been done correctly. In many areas, this technology is wholly inappropriate and often ends up depriving many people of their PDS entitlements. (Right to Food India, 2016a: 24)

The technical problem, remarks again the campaign, is combined with policy issues which are covered, among other sources, by Sinha's (2015) article on the replacement of cash for food. Sinha shifts the attention on to the Jan Dhan Yojana, Aadhaar and Mobile (JAM) Trinity model, discussed in Chapter 5 of this book. The purpose declared in the Economic Survey 2015/2016 (Government of India, 2015) is that of replacing food subsidies under the PDS with cash transfers, which would avoid the distortion associated to subsidies and the room for diversion that the PDS presents. Concerns towards the move have been raised by PDS recipients, as our interview with Adeela in Chapter 2, among others, revealed; to these concerns, Sinha adds the structural issues brought by a cash transfers programme, which is not designed to identify needy beneficiaries and incurs issues of price fluctuation (Sinha, 2015: 18). In exposing the issues related to an Aadhaar-enabled passage to cash transfers, Sinha remarks once again the cruciality of fair practices in relation to the incorporation of digital ID in food security schemes.

These considerations underpin the problematising role that the Right to Food campaign has acquired towards Aadhaar in food security programmes. Exclusions from the PDS have been studied in Chapter 4; but a similar point applies to Anganwadi centres, targeted in June 2022 by the central government's decision to cut funding to states that do not ensure that ICDS users are registered with Aadhaar. The move makes the funding of ICDS – for a programme that provides nutrition to over 79 million children a year – conditional to Aadhaar seeding of Anganwadi user records, posing a double burden on ICDS beneficiaries and on the country's state administrations. For beneficiaries, the risk is that of having key sources of nutrition made conditional to Aadhaar; but for state governments, programme funding is at risk of being cut, which is especially problematic for poorer states and for states more heavily dependent on central government funding (Article 14, 2022). This move (Article 14, 2022) is promoted with the purpose to remove fake beneficiaries from the ICDS, fostering a mobile app to track them; its exclusion risk is however especially worrying, as it concerns categories that are already at nutritional risk.

As a result, while absent from the founding statement, engagement with digital ID is paramount in the work that the Right to Food campaign conducts. On the one hand, positive forms of digitalisation are recognised by the campaign's work; the informative potential of digitised ration card records, as well as electronic weighing scales in ration shops, is framed as a route to virtuous datafication. But when it comes to Aadhaar's centralised architecture, the campaign's concerns abound; the opposition to the Aadhaar-centred move made on ICDS in 2022 crystallises a critique that has formed over the years, illuminating the exclusionary effect of Aadhaar and its connection to policies that can harm recipients. While noting this, the Right to Food campaign has engaged digital systems in proactive forms; a key example is its work towards cash-based maternity entitlements, introduced by the NFSA but unequally implemented across the nation.

Proactive Focus: ID and Advocacy for Maternity Entitlements

Maternity entitlements are an integral component of the NFSA. The law states that pregnant and lactating women (up to six months after childbirth) are entitled to a hot, nutritious meal from the local Anganwadi, and a total of Rs. 6,000 per child for every woman 'not already benefiting from maternity entitlements under other laws or from regular government employment' (Right to Food India, 2016: 10). The campaign's primer 'Anganwadi for All' (Right to Food India, 2016b) substantiates the point that nutrition during pregnancy and lactation is crucial to mother and child's health. ICDS is India's most important scheme for children under six years of age, and its sustenance to mothers has been written into policy since the launch of the scheme in 1975. The Anganwadi primer, along with the activity of the campaign through the years, illuminates the cruciality of defending the right to food in a mother and child health perspective.

As the campaign notes, proper antenatal and postnatal care is needed to break the intergenerational cycle of hunger. Despite progress achieved from measurements taken in 2005–2013/2014, India still has worse malnutrition rates than most countries in the world (Right to Food India, 2016b: 2); malnutrition rates are three times as high among poor children as among the economically better-off. Apart from ICDS, at least two more schemes have an impact on maternity entitlements. The Indira Gandhi Matrutva Sahyog Yojana (IGMSY), which translates as Indira Gandhi Maternity Support Scheme, introduced in 2020, is a conditional cash transfer scheme for pregnant and lactating mothers of 19 years and above for the first two live births, which was brought under the NFSA in 2013. The Janani Suraksha Yojana (JSY), targeting the same group of women, integrates cash assistance with 'antenatal care during the pregnancy period, institutional care during delivery and immediate post-partum period in a health centre by establishing a system of coordinated care by field level health worker' (ILO, 2016). According to Right to Food India (n.d.) 'there is still a long way to go' when it comes to ensuring nutrition for mothers and children, and maternity entitlements are a substantial part of progress made on that way.

The NFSA, the primer continues, affirms women's rights to maternity entitlements, even though 'the minimum amount (Rs. 6,000 per child) is very low' (Right to Food India, 2016a). The point is further substantiated by the campaign:

> Women's need for nutrition, particularly iron, increases sharply during pregnancy. Lack of iron leads to anaemia. According to the National Family Health Survey 2005–6, 58 per cent of Indian women in the age group of 15–49 years are anaemic. Moreover, many Indian women are constrained to do hard manual labour until very close to the end of their pregnancy, and resume work soon after delivery. For these and other reasons, there is an

urgent need to recognise every woman's right to maternity entitlements as well as adequate rest and nutrition during pregnancy. (Right to Food India, 2016a: 16)

It is on mothers' need for nutrition that the cash transfers associated to maternal entitlements can play a key role. The NFSA acts as a partial compensation for loss of wages during pregnancy; it also helps pregnant women have adequate rest, and it contributes to children's right to food by reducing the risk of low birthweight (Right to Food India, 2016a: 16). Maternity entitlements, it should be noted, are universal; the NFSA covers all those women that are not eligible to similar benefits under other schemes (Right to Food India, 2016a: 18), making a targeted scheme effectively universal in its reach to all women in perinatal situations.

The Right to Food campaign, we have seen, has played a highly reactive role towards the incorporation of Aadhaar in anti-poverty schemes. But when it comes to maternity entitlements, the position of the campaign is different; it is indeed one of advocacy, as the NFSA primer (Right to Food India, 2016a) notes that, in 2016, maternal benefits were still missing in all Indian states except for Tamil Nadu and Odisha. In Tamil Nadu, the Dr Muthulakshmi Maternity Assistance Scheme (MAMTA) provides 'Rs. 12,000 per delivery, to be paid in three instalments of Rs. 4,000 each. In Odisha, the MAMTA scheme provides Rs. 5,000 in four instalments' (Right to Food India, 2016a: 18–19). Yet both schemes are restricted to women over the age of 19 and to the first two live births, which underpins the importance of a cash transfer model that guarantees the nutrition standards that the NFSA advocates for.

The Right to Food campaign's advocacy of cash entitlements for mothers unpacks an important point; schemes like IGMSY, JSY and the maternal entitlements of the NFSA require accurate, data-supported identification of beneficiaries. But at the same time, they do not require the subjection to Aadhaar that other schemes, including the PDS, experienced over the years. The campaign offers a key example of how cash transfers, targeted to programmes tailored to needy categories, can make a systematic difference for users; and yet, Aadhaar's biometric architecture is delinked from the benefits that the same transfers are meant to result in. By advocating cash transfers under the NFSA and other schemes, the campaign offers an important example of proactive data activism, one that is further substantiated by its advocacy of ration shop level computerisation to support beneficiaries.

Kenya: New ID, Similar Injustice?

As noted in Chapter 4, Kenya has recently witnessed the pilot phase of a new digital identity system. On 13 May 2023, Kenya's Principal Secretary State Department for Immigration and Citizen Services announced the government's plans for rolling out the Unique Personal Identifier (UPI), also referred to as *Maisha Namba* ('life number'). The system is designed to identify Kenyan citizens 'right from birth', and along the course of their lives; it is meant to be used through citizens' existence 'as ID number, NSSF number and on everything up to death certificate number'.[2] The plan, argued the Principal Secretary, was created to establish digital identity for all citizens, as well as enabling their access to government services through a dedicated digital platform (Kenya News Agency, 2023).

In the same announcement the Principal Secretary remarked the difference between UPI and Huduma Namba, the scheme which, launched in 2019 to provide a digital ID to all Kenyan citizens and foreign residents, was declared illegal by the Kenyan High Court in 2021 (Privacy International, 2022). And still, civil society organisations show sustained concerns about UPI, initially reflected in an open letter published on 24 May 2023 and signed by multiple organisations in the human rights space (Access Now and eight other signatories, 2023). The central problem posed by the letter lies in the danger of repeating, with UPI, the same injustice-producing mistakes that led to the cancellation of Huduma Namba, involving substantial risks for human rights protection:

> Lessons from Huduma Namba rollout show that if not done right, implementing digital ID systems produces further inequalities for minority and historically marginalised communities, including the Nubian, Borana, Swahili and Somali communities as well as double registered persons (Kenyans whose biometrics are in the refugee database) who already struggle with systemic discrimination in obtaining registration and nationality documents. The introduction of UPI could also increase inequities for vulnerable communities who do not have access to birth certificates and IDs due to discrimination, distance, cost, corruption, and other barriers and they may be further excluded from the opportunities provided by UPI. (Access Now and eight other signatories, 2023, n.p.)

The letter's signatories make a specific request to 'slow down' the implementation of UPI. The request echoes the 'not so fast!' plight by the Centre for Human Rights and Global Justice (2022: 79); their report noted the importance, when introducing digital identity systems, to take the time to build an evidence base on potential benefits and injustices. In doing so, the report especially recommended to 'take all

---

[2]The National Social Security Fund (NSSF) is an organisation that offers social protection to all workers in Kenya, receiving members' contribution and paying out benefits (https://www.nssf.or.ke).

necessary steps to anticipate and mitigate possible harms in advance' (2022). The UPI, note the letter's signatories, risks incurring a replication of the same mistakes made with Huduma Namba, and ultimately does not present the traits of the 'human rights-based digital ID' (Mutung'u, 2021) that Kenya needs to build a responsive digital ID system.

In December 2023, in a fashion similar to the Huduma Namba case, the Kenyan High Court has blocked the rollout of Maisha Namba, on grounds of lack of a Data Protection Impact Assessment (DPIA). Abuya (2023) notes how data protection concerns were at the basis of the Court's decision; through Judge John Chigiti, the Court ruled for a halting of the project 'including the digital card, digital ID, unique personal identifier and a National Master Population Register before and without a data protection impact assessment, per section 31 of the Data Protection Act' (cited in Abuya, 2023). Section 31 of the Data Protection Act, notes the Kenyan edition of Nation (2023, cited in Burt, 2023), requires the government to perform a DPIA before processing digital identity data. It is important to note that the halting of the project suspends the implementation of all its components; not just the chip-bearing physical ID card, but also the UPI associated to it and the National Master Population Register meant to underpin the assignation of cards (Abuya, 2023; Burt, 2023).

The announcement of the planned UPI rollout came just ten days before the 2023 edition of ID4Africa, which took place in Nairobi on 23–25 May. The case of UPI shows again how Kenya's debate on digital ID involves an ecosystem of government actors, vendors and human rights organisations; concerns on the treatment of minorities are prominent in the debate, featuring the urge for avoiding reproduction of the historic marginalisation of selected communities (Gonzalez, 2023). As noted in Chapter 4, the colonial production of Kenya's borders has crystallised discriminations that, first reflected in colonial measures of identification, were reproduced in present-day biometric systems (Weitzberg, 2020a; Mutung'u, 2021). The plight of double registered people, whose presence in a refugee database denies them a national ID card, acquires a new valence in the light of the plans for rolling out UPI.

Enacting Infrastructure Justice through Deregistration

As noted in Chapter 4, double registered people find themselves in legal limbo; their presence in the register of refugees, shared by the UNHCR with the Kenyan authorities in 2012, denies them the possibility to obtain a national ID card, which is needed to participate in most aspects of Kenya's civic life (Weitzberg, 2020b; Haki na Sheria, 2021; Mutung'u, 2021). In the light of UPI, a new form of injustice seems in sight for them; the lack of a national ID card (and of the possibility to qualify for it) makes double registered persons ineligible for an UPI ID, risking replication of the situation of factual statelessness in which they are already placed (Access Now, 2023). Addressing the issue of double registration is central to human rights protection in Kenya, and it closely links ID production to the plight of more than 40,000 people (Haki na Sheria, 2021, 2022; Namati, 2021).

The Haki na Sheria initiative is among those upholding the most long-term engagement with double registered people. Based in the border county of Garissa, Haki na Sheria was founded by Kenyan Somali law students at the University of Nairobi in 2010, at a historical time characterised by the country's run-up to the promulgation of the Constitution (Haki na Sheria, n.d.). The founders, notes executive director Yusuf Bashir (Haki na Sheria, n.d.), saw the opportunity to address the historically produced marginalisations and human right violations in the country. First registered as a community-based organisation in Garissa, the organisation became registered with the NGO board in 2017, and centrally addressed – among other topics in the space of civic and environment justice – the plight of residents unable to get a national ID, which resonated with the life histories of the founders.

Such a background informs the role that Haki na Sheria, since the emergence of the issue of double registration, played in defence of the people affected. The Haki na Sheria (2021) report 'Biometric Purgatory: How the double registration of vulnerable Kenyan citizens in the UNHCR database left them at risk of statelessness' powerfully voices such stories, through interviews conducted with people affected by double registration in Garissa county. The report documents both the circumstances that led people to be double registered, and the impact that the lack of a national ID has on their lives; in doing so, it illuminates a plethora of similarities among the stories of double registered persons. A first similarity lies in the reason for registration as refugees; services including food, free education and medical services were made available in the refugee camps, but not for the Kenyan nationals escaping the multiple droughts that affected northern counties (Haki na Sheria, 2021: 4). 'We just wanted a better life' was the plight of many; this led numerous people to head to the refugee camps with their families, registering themselves and their children in the needed pursuit of essential goods (Haki na Sheria, 2021: 27).

As noted in Chapter 4, another common factor across people's stories lies in the lack of awareness of the consequences that, years later, being registered as refugees would have had on their and their children's ability to enjoy citizenship rights. Many discovered their status only when applying for a national ID card, which in Kenya usually happens at the age of 18; at the time of application, they were notified of the presence of their fingerprints in the refugee register, resulting in their applications being turned down. Stories from Haki na Sheria's report cluster strongly around this unawareness, remarking how – had people known the cost that registration would have had in the future – that would have weighted on a decision made for essential, immediately needed food and service provision. Community members interviewed in the report suggested that the UNHCR knew about the issue; a leader declared 'the host community advocated for programmes for their own wellbeing, arguing that the presence of services such as food and education for refugees was creating tension between the refugees and the host community', while another stated that politicians warned the UNHCR that Kenyans were registering as refugees (Haki na Sheria, 2021: 29).

What is central, in double registration, is the infrastructural nature of the issue. Let us remember that the problem is produced by a dynamic – the sharing of the UNHCR register of refugees with Kenyan authorities – that affects the core of the national ID database. It is such a sharing that makes the fingerprints of registered refugees searchable when applying for a national ID card; for its design, the infrastructure recognises the person as a refugee, automatically producing a negative response to an application for Kenyan citizenship (Haki na Sheria, 2021). The dynamic that bars the person from obtaining an identity card is inscribed directly in the body of the technology.

And it is on such an infrastructural nature that the activism of Haki na Sheria, as well as that of the ecosystem of actors supporting victims of double registration (Mutung'u, 2021), impinges to address the problem. What organisations advocate is a dynamic called *deregistration*; this consists in the removal of people's records and fingerprints from the refugee database, so that a new record can be created for them in the register of persons. The problem is infrastructural as it concerns the very way the database is constructed; only by removing the person from a database which erroneously reflects their identity, and results in denial of essential rights, can the person's ability to obtain a national ID card be restored. Deregistration comes across, using the concept coined by Cheesman (2022a), as an act of *infrastructure justice*, where correcting the infrastructural construction of the refugee database results in the ability to restore people's fundamental rights.

It was advocacy for deregistration, as well as multiple forms of litigation and judicial tools in support of double registered people, that informed action by the Kenyan government on the matter. For years, organisations including Haki na Sheria, Namati and the Nubian Rights Forum were at the forefront of this engagement, providing legal aid where very limited support existed. Such work contributed to a landmark achievement in January 2022; the government announced a plan to issue ID cards to 40,000 double registered persons in Garissa, an announcement welcomed by a press release by Haki na Sheria (2022). The press release noted, at the same time, that many more people still need to see their plight addressed; this called the Ministry of Interior to set up a fair mechanism of rights redressal, for the issue of double registration to eventually come to a complete resolution.

#WhyID: A Data Activist Lens in Practice

India's Right to Food campaign and Kenya's movement for double registered people are important stories of data activism, both focused on issues that emerge at the level of national perspectives. For the Right to Food campaign, a central issue are the negative effects of digital ID on people's access to food security schemes, which are combated by routes to access that can be technologically mediated. For double registered people, denial of citizenship rights stems from the very construction of the national register of persons, which, in virtue of interoperability with the UNHCR database, classifies them as refugees. Activist advocacy therefore pertains to deregistration from the refugee database, which is instrumental to the plight for fulfilment of their right to citizenship.

But campaigning for restoration of rights lost through digital ID systems also presents transcontextual dimensions, involving actors from diverse realities. Over the years, multiple international symposia have emerged in relation to the perils of unfair ID. The work of Privacy International, engaged since 1990 in promoting the right to privacy across countries, is paradigmatic of this international spectrum; across its lifetime, the organisation's engagement with ID-enabled infringement of rights, biometric mass surveillance and its dangers for human security have generated wide actor networks of impact (Privacy International, n.d., 2019a, 2019b). A central feature of such impact is the proactive nature of the networks' advocacy, committed, as previously discussed in this book, to imagining routes to fairness beyond the harms that the organisation engages. Another example is provided by the civil society coalition that emerged in response to Kenya's proposed Maisha Namba, and which worded its contention as follows:

> We first address our concerns about the government's failure to engage citizens and allow for meaningful public participation (. . .) a revised national ID can be of significant benefit if it is designed in a way that is responsive to public concerns, protects privacy, and helps eliminate (not replicate) the discrimination and inequity inherent in the existing system. (Haki na Sheria, 2023, n.p.)

Multiple examples of such aggregative works are found in today's response to ID unfairness. One such case is #WhyID, a campaign that, coordinated by Access Now, groups entities across nations, contexts and organisational types. The campaign's genesis reflects its heterogeneous nature; it was a set of 'civil society organisations, technologists, and experts who work on digital identity developments' to first gather through digital means, expressing concern about harms of 'ill-considered, badly designed, and poorly implemented digital identity programmes' (Access Now, n.d.). The campaign started with an open letter to 'leaders of international development banks, the United Nations, international aid organisations, funding agencies, and national governments'; its purpose was to denounce the severe harms inflicted by centralised ID programmes on people, and at the same time, imagine routes for such harms to be detected and avoided. Grouping together voices from different provenances, the campaign clustered around the fundamental question on *why ID*, and *what* is wrong with 'centralised, ubiquitous, data-heavy' forms of digital identification? (Access Now, 2021: 3).

The question, the open letter continues, stems from the gap between the proven real harms, and the scantly documented promised benefits of centralised ID programmes, a finding recently confirmed by the Centre for Human Rights and Global Justice (2022: 8–9). Harms, it is argued (Access Now, n.d.), are made scalable by the scaling of digital ID schemes; in addition, several so-called 'developed' countries scrapped identity schemes of the same type, which are instead being increasingly rolled out in less wealthy nations. A case in point is the identity scheme proposed by the UK government in 2004; at a time in which no such programme existed worldwide, the scheme proposed the coining of ID cards linked to a central database, holding confidential personal data and face, eye and fingerprint scans (Whitley & Hosein, 2010). Met with concerns raised by the civil society, the programme was exposed by multiple entities, including the LSE Identity Project, in terms of its costs and privacy implications, and eventually scrapped in the year 2010 (Whitley & Hosein, 2010). Analogous initiatives, notes the #WhyID campaign, are instead financially pushed across less wealthy countries, with limited consideration of evidence around their harmful effects (Access Now, n.d.).

Central to the #WhyID message, articulated across the years by multiple organisational realities (cf. Privacy International, 2019a), is the notion that 'human agency and choice form the foundation of human dignity' (Access Now, n.d.). This is why the 'why ID?' question is so crucial; it is important that all those affected by digital ID are informed, first of all, of why they are being subjected to digital identification, before even conveying the 'who, when, where and what' answers on the programmes affecting them. The #WhyID letter signatories range from privates to organisations operating in many spaces of human rights defence; grouping such entities is a set of shared ideas, questioning centralised ID programmes and exposing the effects they can exert on people. The same advocacy resulted in the request, in another open letter, for the World Bank and associated actors to cease the implementation of digital ID programmes that are proven harmful for human rights (Access Now and 45 other signatories, 2022).

The transnationality and transcontextuality of such requests have characterised the #WhyID campaign since its beginning. Said with Milan and Van der Velden (2016), the campaign's activism has clustered around reactive and proactive components; its reactive building blocks centre around the notion of 'Big ID', first proposed by Access Now (2021) in the report 'Busting the Dangerous Myths of Big ID Programs: Cautionary Lessons from India'. Centred on India's experience with Aadhaar as a globally relevant foundational ID scheme, the report builds on the dichotomy between Aadhaar's promises – of a unique, access-facilitating entry point for India's social programmes – and a reality characterised by exclusions of entitled users and undue surveillance, with the incoming danger of enactment of a surveillance state (Access Now, 2021: 13–14). It is in relation to Aadhaar that the notion of Big ID is defined as 'large programmes, linked to the public sector, which seek to assign citizens and residents an ID stored in a centralised database, linked to biometric authentication' (Access Now, n.d.).

Big ID, notes the report in point, has been associated to a set of 'myths' that, well-embodied by India's experience with Aadhaar, do not correspond to the realities lived by digitally identified people. The report articulates 12 such myths, the first of which is on how Big ID would be needed to give people a legal identity; this point makes digital ID a fundamental condition for being 'seen' by the state as it performs its fundamental functions. But the equation breaks down as people's visibility – while it can be technologically enabled – remains, the report argues, a political choice; as noted in this book, while technology can crystallise policy decisions, it does not alone reverse them, or restore justice where fundamental rights are being denied. Kenya's history of double registration again exemplifies this; people's right to be 'seen' as citizens is not denied by the technology alone, but by the rationale that qualifies all double registered people as refugees. Big ID, argues Access Now (2021: 3–4), lies at the root of such injustices, crystallising patterns of harm generation that an uncontrolled reproduction of Aadhaar-like programmes over countries risks to scale up.

While the reactive component is strong, *proactive* data activism is equally central to the work of the #WhyID campaign. It is crucial to understand that busting the myths of Big ID, as in Access Now (2021), is not a route to ID nihilism; it is, on the contrary, a fundamental step in proposing principles that – rather than denying the rights of users of digital ID programmes – assert such rights in the design and implementation of ID schemes. Essential to the advancement of such rights is the #WhyID response to the World Bank Principles on Identification for Sustainable Development, translated in actionable principles by Access Now (2020). In a paper called 'National Digital Identity Programmes: What's Next?', the organisation lays out principles under three categories: 'Data Protection and Privacy', 'Governance' and 'Cybersecurity', all founded on the recognition of harms induced by digital ID. It is arguably on the basis of such principles that more just forms of ID can be imagined.

The proactivity of the principles laid out by Access Now (2020) is central to understanding what a 'fair ID', consciously tackling the harmful features of what we termed unfair ID systems, may look like. These principles emerge in response to the 'Principles on Identification for Sustainable Development', facilitated by the World Bank's ID4D initiative; the principles (ID4D, 2019) give guidelines for designing and implementing the ID systems that this book has been dealing with. These principles are endorsed by organisations that, especially in less wealthy contexts, fund and promote digital ID. The categories of 'Data Protection and Privacy', 'Governance' and 'Cybersecurity' come from intersecting identity principles with the defence of human rights, a point which is further elaborated in Chapter 8.

These principles show that, in denouncing the outcomes of unfair ID, a proactive route can be adopted with actionable implications. Signatories to the principles are called to make key commitments; they are bound to refrain from engaging ID schemes that do not adhere to the principles, and to restrict funding and support for those programmes (Access Now, 2020: 24). In addition, the signatories are called to institute mechanisms for civil society organisations to provide feedback from the projects they support and fund. ID fairness may come, in other words, from reimagining the politics of the artefact from the eyes of beneficiaries.

Data Activism: Towards Fair ID?

This book has studied digital ID systems through a *data justice* lens, substantiating the notion of unfair ID and illuminating the shapes it can take for people. At the same time, a *data activism* lens has been crucial in bringing up narratives of digital ID resistance and solidarity, of which the stories from this chapter constitute prominent instantiations. Inspiring such a lens is again the extended notion of data activism from Milan and Van der Velden (2016) as 'the range of sociotechnical practices that interrogate the fundamental paradigm shift brought about by datafication'. It is such practices that reflect the forms of sociotechnical engagement in which people partake, acting *on* data (for instance by contesting unfair ID practices) and *through* data, leveraging digital systems in the making of contestation.

All three stories from this chapter show the coexistence of *reactive* forms of data activism, which contest unfair data practices, and *proactive* forms that support the defence of human rights. In India's Right to Food campaign, fair ID practices are seen as a gateway to essential food supplies; the initiative denounces the unfair outcomes of Aadhaar, but it does so *through* the advocacy of alternatives, such as digital ration card records and electronic weighing machines set in ration shops.

This advances fair, user-centric ways of doing ID; for ration card records, digitality means the ability for people to check their own household data, requesting changes, additions and deletions of members wherever needed. Electronic weighing scales offer a route to transparency of ration shop transactions; assuming proper functioning of the system, they reveal the quantity of commodities sold to users, providing them with essential information on the goods sold in each transaction.

In Kenya's movement for deregistration of double registered people, reactive data activism is centred on denial of people's citizenship rights on the grounds of their presence in the refugee database. Using Milan and Van der Velden's (2016) lexicon, reaction is against a database that does not discriminate in relation to people's rights, but only in relation to how their records are classified. And even here, a proactive component emerges; the advocacy of deregistration pushes towards a virtuous practice, that of deleting double registered records so to enable people to receive an ID. It is a case where proactivity and reactivity are mutually responsive; only by denouncing unfair data practices, i.e. denial of an ID to victims of double registration, can the virtuous practice of deregistration of victims be legally supported and allowed to take place.

Another connubium of reactive and proactive data activism is offered by the #WhyID campaign. The genesis of the campaign tells its duality; born to denounce the harms linked to 'centralised, ubiquitous, data-heavy' digital ID systems (Access Now, 2021: 3), it was also born by means of digital aggregation, which connected multiple realities in challenging harmful forms of ID. As epitomised in the *Digital ID Toolkit*, the most recent instrument deployed by Access Now (n.d.) in the scope of the #WhyID campaign, there are multiple routes to mitigating ID harm; these impinge on a minimalistic approach to ID, which captures the minimum possible data points to achieve the best balance of data capture and human rights protection. The digital ID toolkit is in effect produced to support understanding of how alternative, rights-protecting visions of ID can be produced.

Summary

Many more stories of ID resistance exist beyond this chapter. And the chapter has, at the same time, brought the message that ID resistance takes multiple shapes. Reactive forms – which contest data practices on grounds of unfairness – are met by proactive forms, which leverage digital systems to build mechanisms of advocacy and solidarity across borders and contexts. Reactive data activism gives us a clear picture of shapes of unfair ID; but it is proactivity, with its alternative possibilities, that gives a future to ID justice. And it is on proactivity that we build in Chapter 8, venturing into the imagination of fair ID.

8
IMAGINING FAIR ID

This chapter leverages the learnings from the book, in terms of both injustice and resistance as related to digital identity. It does so with the goal of learning from injustice to imagine forms of *fair ID* that, put into practice, have the potential to effectively combat digital ID-induced harm. For this purpose, I put forward the intent of *reversing dark matter*, a notion that impinges on the idea of dark matter defined – as in Chapter 6 – as the core features of digital ID resulting into harm. To reverse dark matter, and embed virtuous features in the making of ID systems, the idea of *infrastructure justice* (Cheesman, 2022a) acts as a guiding light for collective efforts, further corroborated by other concepts derived from research and activism on digital ID.

Smart Cards in Tamil Nadu: Building ID Fairness?

In 2017, the Indian state of Tamil Nadu, similar to other states, introduced biometric recognition in the delivery of PDS goods in ration shops. The system introduced in Tamil Nadu differed, however, from the Aadhaar-Based Biometric Authentication (ABBA) described in Chapter 4, where the collection of monthly rations was enabled by fingerprint recognition. The biometric PDS of Tamil Nadu is based on smart cards to be produced at the ration shop. Each smart card, equipped with a QR code, is associated to the households' details, and to a phone number for notification of delivery of goods. When producing the smart card in the ration shop, the point-of-sale machine screen displays the household data, as well as whether, and in which proportion, the monthly ration has been collected. If the ration, or part of it, has not been collected, the machine prints out a paper with user details and quantity distributed; the user then joins a second queue, where the ration is disbursed to them (Hundal et al., 2020: 3; Carswell & De Neve, 2022: 130–131).

The smart card of the Tamil Nadu PDS constitutes one of those innovations that, while still Aadhaar-based (predicated on registration of user data in the Aadhaar database), offers an alternative to ABBA's architecture. In a two-state case study of Karnataka and Tamil Nadu, Hundal et al. (2020) note the differences between the two systems. Under ABBA in Karnataka, the person is required to input their fingerprint in the scanner, with ration delivery being conditional to fingerprint recognition. In the Tamil Nadu system, the person's bodily markers play no authentication role; what the point-of-sale machine scans is the QR code on the smart card, associated to the households' credentials. The smart card reader, note Grace Carswell and Geert De Neve – respectively, Professor in Geography and International Development and Professor of Social Anthropology and South Asian Studies at the University of Sussex – shows on a screen the household's monthly allocation, as well as the balance of commodities that are still to be collected in that month (Carswell & De Neve, 2022: 131).

It is arguably the lack of fingerprinting, at least in the system's original version, that led to the idea that Tamil Nadu's biometric PDS could offer a better authentication solution than ABBA.[1] In an article titled 'Smarter than Aadhaar', Khera (2018) illuminates the improvement brought by smart cards with respect to fingerprint recognition in the PDS. The smart card, Khera notes, keeps a digital trail of all transactions for a given ration card; this is crucial when having to retrieve a household's card history, and the transactions that took place over time. Secondly, the presence of a QR code on the smart card delinks ration delivery from the person's bodily features. Khera (2018) notes how this brings important advantages with respect to ABBA, as it reduces likelihood of fingerprint misrecognition and enables immobile persons, such as the elderly, to have their rations collected on their behalf. Both arguments seemingly make the smart card 'smarter' than ABBA, which requires, in the words of Drèze (cited in The Wire, 2016), 'multiple fragile technologies' to work at the same time.

From the point of view of the politics of the artefact, the smart card of Tamil Nadu presents important elements to potentially enable fairness. Both of Khera's (2018) observations can be seen in this logic; the artefact, a smart card with an embedded QR code, is functional to producing the digital trail of transactions associated to each household. This information is especially precious as, when a person comes for collecting the monthly quota, the ration dealer needs to know if the quota has already been lifted; a transaction can therefore not be duplicated, and monthly allocations are in principle ensured. In addition, the digital system displays the balance of goods still available for households. It makes it possible, for instance, to collect the allocated quota in different portions, making sure that the correct monthly balance will be fulfilled. With the scanner's screen being visible both to the ration dealer and to users, the artefact seems to be designed for transparency.

---

[1]Conversation with Carswell and De Neve who visited ration shops in rural Tamil Nadu in 2022 revealed, however, that by that time the smart card readers had been integrated with a fingerprint reader as well.

The delinking of ration collection from fingerprint recognition also constitutes a substantial advance in terms of tackling the design injustice issues discussed in Chapter 6. The ABBA system of Karnataka, note again Hundal et al. (2020), incurs the risk of not recognising fingerprints of genuinely entitled users; after the third failed attempt to authenticate, people are invited to step aside, which leads to the coping mechanisms (household members queuing together to maximise chances of recognition) described in Chapter 4. In addition, the artefact 'poses' the condition, enacted by fingerprint-based authentication, that the entitled person needs to be present at the time of delivery. This is again difficult for immobile persons (Khera, 2018), and can put into serious predicament their possibility of collecting rations. Replacing fingerprint authentication with a QR code, the design injustice that excludes non-readable bodies from fair delivery of goods seems to be fixed at the basis, by delinking a person's bodily features from their ability to authenticate.

The logic embedded in Tamil Nadu's smart cards seems indeed one of combating ID injustice. The biometric PDS in the state, it should be noted, has given up the use of paper. The confirmation of ration delivery happens through a text message, automatically sent to the number connected to the ration card when the transaction is completed. The ethnography of the PDS by Carswell and De Neve (2022) teases out the promises embedded in the system: 'transparency', connected to a technologically revamped architecture of ration disbursement, is translated in the new functioning of the ration shop interface. On the one hand, it is that interface that technology seeks to transform, by inducing visibility in the previously opaque mechanism of ration disbursement. But on the other, it is within the ration shop interface that problems arise, producing, as shown in Carswell and De Neve (2022), opacities and information gaps with respect to the previous system.

The Return of Informational Injustice

The work by Carswell & De Neve (2022) takes the perspective of digital identity users, researched at the interface with service providers. In this case the interface is again the ration shop, where the physical embodiments of the state, such as ration dealers, are encountered. With their long-term work in the Tiruppur area of Tamil Nadu, Carswell and De Neve uncover stories from two villages where the new system has been implemented. Narratives from users recount the artefact beyond its politics, illuminating how promises of transparency and fairness are translated in the technology through which rations are accessed.

It is, to begin with, the promise of 'transparency' that is problematised by Carswell and De Neve's (2022) informants. One of their biggest challenges is the digital nature of the receipts for collected rations. Before the biometric system, the ration card consisted of a paper booklet, where the ration dealer would write down the type and quantity of goods collected. While the booklets have been replaced by smart cards, collection receipts have been replaced by text messages displaying, in English, the collected goods and quantities. The text message, sent to the phone number that the ration card is registered with, automatises the conveyed information: this is meant to reduce the risk of 'tampering' with quantities that, note Hundal et al. (2020), is instead present in ABBA, where the weighing machine for disbursement is physically disconnected from the point-of-sale machine.

The promise of transparency is met, however, with multiple recounts of frustrating experiences by users. To begin with, text messages are in English; since many recipients do not speak this language, some users end up deleting the messages, or get them checked by a relative or neighbour with some knowledge of English (Carswell & De Neve, 2022: 132). On the one hand, the automatic text message notification should remove the ration dealer's possibility to tamper with the quotas of goods sold; but on the other, due to use of English, it drastically reduces users' ability to interpret such information. In a way paradoxically, note Carswell and De Neve (2022: 132), text messages end up adding a further layer of intermediation to the system that consists of the relative or neighbours' help to interpret the text messages, whose information on collected quotas – and on the balance of monthly goods remaining – is crucial for future purchases.

Secondly, the text message with receipt of transaction is not necessarily sent to the phone of the person completing the transaction, but to the phone number linked to the smart card. Reminiscent of discussion of gender and age discrepancies in mobile phone ownership (Chaudhuri, 2021), Carswell and De Neve (2022: 133) note how such discrepancy affects especially women, whose smart cards are typically linked to the phone of a male relative – be it a husband, father, son or an in-law. As noted at multiple points of this book, gender plays a substantial role in the discussion of ID fairness; in the case of ration shops it intersects with age, as older women, showing up to collect their rations, will rarely receive the confirmation message on their own phone (and if so, many of them will be unable to interpret it). Informants report of messages not going through, or going through much later (Carswell & De Neve, 2022: 131–132); partially as a result, they unanimously report their preference for the ration card paper booklet, where transactions were immediately legible (Carswell & De Neve, 2022: 131).

But a deeper, infrastructural problem comes up when discussing the Tamil Nadu case. This has to do with the material, paper-based nature of the artefact (Rao, 2017; Carswell & De Neve, 2020): while paper-based transactions are blamed with 'tamperability' by outsiders, from users' perspective they offer the secure materiality provided by a written proof of transaction (Carswell & De Neve, 2022). Informants detail how with the ration booklet, they could show up at the ration shop and show – booklet in hand – the month's remaining balance; this cannot be done with a text message, as the absence of a received text cannot be taken as proof of balance.

Several incidents (Carswell & De Neve, 2022: 130–131) are recounted where users did show up to the shop, to find the ration dealer claiming that their monthly ration had already been collected; when the user claimed not having received the text message, ration dealers just blamed it to the system being down, or to network delays. In the opposite fashion of the artefact's politics, aimed at increasing transparency in the PDS, the system took bargaining power away from users, introducing novel information gaps (Carswell & De Neve, 2022: 131).

What is starking in the story of the Tamil Nadu PDS is the open discrepancy between a promise of fairness, substantiated in the transition from booklets to smart cards, and the new reality of opacity with which users are confronted. On the one hand, this is reminiscent of the informational injustice studied in Chapter 5, and in particular of the iris scanners that, in Jordan, took away information-rich paper receipts from women refugees (Cheesman, 2022a, 2022c). On the other, it faces us with one of the main dilemmas in the design of fair ID; that of reflecting the artefacts' politics in the experience of recipients, and in their access to their entitlements once the system is in place. The Tamil Nadu smart cards were designed with a vision of transparency; but as revealed in Carswell and De Neve (2022), that vision ultimately did not meet the lifeworld of recipients, and instead resulted in informational injustices that were not present in the paper-based system. It is to navigate such dilemmas, and their relevance to imagining fair ID that the notion of *infrastructure justice* as theorised in Cheesman (2022a) is introduced.

The Lens of Infrastructure Justice

Let us go back to Cheesman's ethnography of blockchain in humanitarianism, discussed in Chapter 5. Blockchain, Cheesman (2022a: 61) notes, is part of what Madianou (2019) refers to as the 'biometric assemblage'; a technological connubium that also comprises of biometrics and artificial intelligence (AI), which shapes the treatment of refugee data from the UNHCR and other entities. The biometric assemblage, argues Madianou (2019: 581), accentuates asymmetries between recipient refugees and humanitarian agencies, illuminating the perils implicit in the wave of biometric refugee registration that started in the 2000s. In the biometric assemblage the blockchain emerges as a promise of positive change, a toolbox to improve refugee identification while at the same time tackling the ineffectiveness of data-heavy systems (Cunha et al., 2020). The promises of blockchain, notes again Cheesman (2022a, 2022b), are especially connected to its decentralised architecture, which offers an alternative to the perverse effects of ID centralisation (Access Now, 2021; The Engine Room, 2023).

A key point in Cheesman's theorisation is in the concept of blockchain as *sociotechnical infrastructure*, where infrastructures are 'systems that connect, enable and sustain social action, underpinning the production and circulation of people and things, resources and capital, knowledge and ideas' (Larkin, 2013; Star, 1999; cited in Cheesman, 2022a: 53). So conceived, sociotechnical infrastructures reach beyond the physical fabric of the artefact. Their components meet the affective, emotional reactions of the people involved, their lived realities and the physical and transcendent lifeworlds in which these are produced. In humanitarian blockchain, the technology's distributed ledger encounters the lived reality of refugees and the needs expressed by their physical and spatial condition. On the one hand, notes Cheesman (2022a, 2022c), the ledger itself is invisible to users, and not encountered in their lived experiences. But on the other, the technology's embodiment is very stark and powerful, and translates in the iris scanning tools that, as noted in Chapter 5, mediate the collection of cash in information-erasing ways, which generate anxiety and frustration in the refugees she interviewed.

It is from the notion of blockchain as sociotechnical infrastructure that Cheesman (2022a) derives the idea of *infrastructure justice*. With this term, Cheesman (2022a: 20) means 'the uneven benefits which sociotechnical systems instantiate', relying on Star's (1999) idea of infrastructure as 'connective and enabling' at the same time. Infusing questions of justice in sociotechnical infrastructures, Cheesman casts doubt on the idea of embedded social good that a blockchain-for-humanitarianism technology embodies in itself. The concept of infrastructure justice takes us to the lifeworlds of women refugees who, having been inscribed into a blockchain-mediated programme, have had their salary payments committed to an eye-scanning technology disbursing not cash, but paper receipts providing only some information on the due payments. Cheesman's concept of infrastructure justice can be seen as a route for study of the biometric assemblage, offering important conceptual tools to understand initiatives and artefacts aimed to realise fair ID.

As noted in Chapter 5, Cheesman's notion of infrastructure justice relies on three interconnected elements: *subjectivities*, *timescapes* and *materialities* (Cheesman, 2022a: 20). *Subjectivities* refer to 'the range of affective, embodied and culturally situated rationales and responses' connected with the infrastructure (Cheesman, 2022a); in the humanitarian blockchain, subjectivities have to do with how refugees experience it at the point of encounter, as it happens with the EyePay machine that disburses payment receipts based on iris scans. But if the focus shifts to other components of the biometric assemblage, such as the smart cards of the Tamil Nadu PDS, then other subjectivities might emerge; for instance, those of the PDS users who do not get a text message upon ration delivery, or get it in a language (English) that requires them to resort to a mediator for understanding their content. Their subjective reality is affected by frustration, especially if compared to the paper-based system that was in force beforehand; while the ration card booklet afforded to claim goods from the ration dealer, the digital system does not, and its design does not meet the need for critical information that they experience.

By *timescapes*, Cheesman (2022a: 20) means 'the temporal and spatial regimes blockchain disrupts and co-creates'. Infrastructures, Cheesman notes, mediate the very construction of time and space; for women refugees in the cash-for-work programme, the time of information reception is delayed by EyePay, whose technology produces receipts that do not specify the days and hours that each person has worked. The same notion points to the space where users do the authentication procedure; the EyePay system moves this to the camp's supermarket, detaching humanitarian action from the locus of salary collection. If applied to the smartcards of Tamil Nadu, the notion of timescapes illuminates the time difference between collection of rations and reception of the text message; it also illuminates the many factors that delay the reception of text messages on ration delivery, from genuine delivery failures to the fact that another person may own the phone to which a ration card is connected.

By *materialities*, Cheesman (2022a) means 'the material practices, processes and architectures that are established and maintained with blockchain infrastructures'. This book started from the notion that, when studying people's encounters with social protection systems, accessing their *loci* of encounter with providers is crucial to understanding their image formation processes. The point is also made by Cheesman (2022a), with her focus on the material practices with which women protect their paper-based EyePay receipts (e.g., folding them in their bra to prevent the ink from rubbing off). A similar case is found in the practices mediated by Tamil Nadu smartcards, where the lack of a material proof of delivery on a booklet offered much greater bargaining power than the non-demonstrable absence of a text message received on their phone.

Subjectivities, timescapes and materialities inform Cheesman's (2022a) notion of information justice, shaping her respondents' experience of humanitarian blockchain. The subjective nature of person–technology encounters makes it difficult to generalise on such interactions, and at the same time, Cheesman's (2022a) concept offers important dimensions to study them. I use infrastructure justice as a conceptual device to begin our guided imagination of fair ID.

Imagining Infrastructure Justice: Reconstructing the User–Provider Interface

We started this book by studying digital identity in spaces where people encounter service providers, giving material flesh to 'the state' or humanitarian agencies. This lens echoes the recent, important book by Rajesh Veeraraghavan, an Associate Professor of Science, Technology and International Affairs at Georgetown

University. Veeraraghavan's (2021) ethnography of India's National Employment Rural Guarantee Act (NREGA) in the state of Andhra Pradesh illuminates how people's understanding of social protection is developed at the last mile, where the person enters in direct contact with the provider and its technologies. It is here that the process of 'patching development', theorised by Veeraraghavan (2021), occurs; it is a process made of localised, iterative dynamics, in which information is progressively made visible to vulnerable users, who are involved in assessment of the scheme through social audits. Rather than in the upper bureaucratic layers where decisions on the scheme are made, it is at the last mile, where people experience the scheme that change occurs.

Similarly, in the PDS ration shops, core understandings of 'the state' are built through physical encounters between the user and the ration dealer. People's conception of digital ID is constructed through these encounters; stories like Aisha's, whose frustration begins from denial of a ration card at a government office, or Ayanka's, whose fingerprint is not read by the ration shop's biometric reader, portray the material spaces of technologically mediated state–citizen encounters. It is in these material spaces that people form images of digital identity, and that they live its reality.

Such philosophy inspired my early study of telecentres in Kerala, conducted in the year 2009 (Masiero, 2011). The term *telecentres* indicates 'shared spaces where technologies such as computers, phones and Internet-connected devices can be accessed', usually for a small fee (Roman & Colle, 2001). Since the 1990s, telecentres played out as a component of a development model aimed at bridging the digital divide; the model soon came to Kerala, a state whose relatively low GDP per capita was matched by an active political life (Isaac & Tharakan, 1995). The birth of *e-governance*, a concept capturing the advent of digitality in public governance mechanisms, reflected this focus. In 2002, the Kerala State Assembly decided for a strategy that was openly inclusive of the economically lower classes, and that leveraged the philosophy of telecentres to turn such inclusivity into practice.

This philosophy inspired the launch of the Akshaya telecentre project, a partnership between the public sector and private actors, in Kerala in April 2002. The assembly of Malappuram district – a relatively resource-poor area of northern Kerala – brought up the idea of starting a public–private partnership to increase e-literacy in the state; the public part was the state government, while the private component would have been played by local community members acting as *entrepreneurs* for managing telecentres. The district government of Malappuram took the idea forward, and mobilised local citizens as entrepreneurs to start up e-kiosk facilities where a local e-literacy course would be imparted (Madon, 2005: 470). As I learnt back in 2009, many community members applied to the Akshaya Project Office to become telecentre entrepreneurs. The screening process for the numerous applicants was tight, and aimed to select highly trusted community members, to take users through the novelty of information technology.

Born as a district initiative, the Akshaya project fast grew to make telecentres the place where people would access the main government services in the state. The model was especially friendly to users who, due to social or economic constraints, did not previously have the chance to access e-services. In an era that preceded the mobile boom, accessing an e-literacy programme was the fundamental premise to make e-governance thinkable (Madon, 2005; Gopakumar, 2007). The e-literacy phase of the Akshaya project involved active mobilisation of the locally selected entrepreneurs; many went door-to-door advertising the new e-literacy programme, to ensure that at least one member for each household turned up to attend the e-literacy course. With 100% e-literacy rates reportedly achieved in Malappuram district, the programme set itself for scaling, leading to Akshaya becoming a state-wide initiative after a two-phase scale-up between 2007 and 2008 (Kuriyan & Ray, 2009).

My study of Akshaya telecentres was conducted months before the launch of Aadhaar in the same year, and about a year before ration card applications in Kerala became viable to be made online. Still, it revealed points that illuminated the reconstruction that telecentres were trying to operate. Stories of telecentre entrepreneurs spoke, first and foremost, of the importance of mobilising trust. A woman entrepreneur in her early 30s told me about the process of going door-to-door in the village, with the specific idea of enrolling women in the e-literacy project; women that, as the e-literacy phase ended, came back frequently to the telecentre for e-services. In a rural area, young entrepreneur Ayla told me of how she started using her telecentre to impart courses of typewriting, which women looking for jobs had found especially useful for a qualification.

But more field stories told me what people derived from the telecentre model. As I discovered in rural areas, farming communities had started using telecentres for meetings aimed at knowledge sharing; a middle-aged entrepreneur, Rajeev, told me about how his telecentre, attended by many people in the catchment area, had become the meeting place for a Bhoomi club, a club of farmers exchanging information on agricultural practices and crop patterns. Providing context-based services was another route seen as successful; some entrepreneurs specialised in services of high relevance to the local community, like a young male entrepreneur who started up courses of Arabic typewriting, especially useful to the many people aiming to migrate to the Middle East from the area. With services tailored carefully for the context and civil society involvement, the goal of *reconstructing the key encounter* between citizens and the state – bringing it from unfamiliar public offices, to friendly and humanly made telecentre spaces – arose as a key component of the Akshaya initiative.

It is the Akshaya story that helps us understand a crucial evolution of the Ration Card Management System (RCMS) that, as noted in Chapter 2, came before Aadhaar in Kerala. Such a system moved the process of ration card management from physical offices to an online interface. On the one hand, the card collection phase remained physical, meaning people had to physically turn up at the Taluk Supply Office to obtain the card. But what became digital was the application phase; this meant people could visit the local telecentre to apply for a ration card, availing entrepreneurs' active help in the process of filling in the online form. In addition, the entrepreneur would help users with the most important operations of ration card management; changing address, addition and deletion of members from and to cards. This increasingly brought people to see entrepreneurs as key service providers, trusting them for key requests in terms of social protection and public services (Madon, 2005; Gopakumar, 2007).

The story of Akshaya telecentres can be read through Cheesman's notion of infrastructure justice. The dimensions of subjectivities, timescapes and materialities are key in the Akshaya experience. People, like Aisha who queued for days in the vain hope to obtain a ration card, would meet the state with stealth and violence (Corbridge et al., 2005: 2), in encounters that the closely knit atmosphere of telecentres sought to replace. The concept aimed to alter the *timescape* of state–citizen encounters, from a space of domination of the service provider to one of participative engagement, where the entrepreneur facilitates complex operations. It can be argued, on this basis, that a central aspect of fair ID is the *materiality* of the encounter between user and provider. Akshaya centres were designed, as the former coordinator of the project told me, to give a human face to public offices, mediating aspects just as essential as the application for a ration card.

Doing Infrastructure Justice: Reversing Dark Matter

The above reflection reminds us of how, as stories from this book have shown, people encounter digital identity *at the interface* with the service provider. It is at that interface that experiences are produced, and that the person forms their own image of the provider and its services. By moving the ration card interface to Akshaya centres, the state of Kerala rebuilt the key *locus* where the experience of digital identity is produced; in telecentres it is now mediated by the village entrepreneur, who has been carefully selected among trusted members of the community. The person's experience of digital identity can then transition from the frustration of being 'unseen' by the state in long queues, bad treatment and unresponsiveness, to an experience where the user is seen and dignified in their interaction with the provider.

The ethnography of Carswell and De Neve (2022) validates this point. Faced with the difficulty of applying for a ration card, especially in the absence of supporting documents, some of their informants became aware of the possibility to apply online; initially in disbelief, some of them eventually managed to complete an online application. The authors themselves learnt how this worked as they assisted a middle-aged woman in applying online and receiving her first ration card ever in this way (Carswell & De Neve, 2022: 135–136). Carswell and De Neve's work mirrors how online applications sought to replace an old locus of frustration with one of support, though there are limitations. For instance, the woman that the researchers helped obtain a ration card online was only awarded a so-called *sugar card*, which in Tamil Nadu is associated to provision of sugar rather than food grains. Yet this enabled obtaining the first instantiation of a ration card for a person, and her household, till then deprived of a basic document for accessing key food security provisions.

But if we go back to the core of infrastructure justice, the issue appears to lie beyond the user–provider interface. Cheesman points to 'the uneven benefits that sociotechnical systems instantiate'; and sociotechnical systems do not instantiate unevenness (or even justice and fairness) through peripheral elements, such as the access points that telecentres provide for ration card applications. Let us go back to the notion of *dark matter*, explored in Chapter 6 in relation to design justice; when digital ID systems are designed with unfair features, the problem may lie in the core of the technology rather than a peripheral, incidental 'dark side'. By the same token, the reconstruction of the *loci* of access to digital identity has limited meaning if the essence of ID technologies is one of dark matter, where harmful outcomes are directly designed in the body of digital ID systems.

As such, the reconstruction of the loci of image formation seems to fix only one aspect of the problem, rather than addressing the heart of design injustice as theorised in Costanza-Chock (2020). This does not mean that shared ICT access, achieved at least in part through initiatives such as the Akshaya telecentre project, has not been meaningful; many works note the impact of telecentres, especially for lower income and otherwise marginalised users (cf. Madon, 2005; Pal et al., 2006; Gopakumar, 2007; Kuriyan & Ray, 2009). At the same time, the idea of *dark matter* invites us to reflect on the point that, if change is invited, it is the heart of the artefact that should experience the change, designing a path to infrastructure justice as Cheesman conceives it. It is to the idea of *reversing dark matter*, by designing virtuous features directly into digital ID artefacts, that we now turn.

Reversal 1: Human Rights Impact Assessments

The notion of *dark matter*, which I have proposed as antithetic to the hegemonic idea of a dark side of IT, has a simple central message. It is the idea that the harm-generating problem lies in the core of a given technology; again this contradicts a long-term focus on the *dark side*, which conceived issues of injustice as unintended and peripheral. In the Information Systems field, the idea of a dark side of IT has created a research agenda defined by Tarafdar et al. (2015) as 'a broad collection of "negative" phenomena that are associated with the use of IT and that have the potential to infringe the wellbeing of individuals, organizations and societies'. The idea is that technology, made for the achievement of societally and organisationally beneficial purposes, had a distorted 'side' that could cause harm; the side is in its own virtue peripheral, and unrelated to the central properties of technology. This inspired a research agenda on the multiple negative 'side effects' of technology.

But the notion of design justice, as in Costanza-Chock (2020), has highlighted the limits of the idea of a 'dark side' of IT. As argued in her book, what generated harm in technology was not just a side, but the very design that made technology biased against already oppressed groups; this made technology unable to meet the principles that the Allied Media Conference (2015) framed as constitutive of design justice. Surveillance studies research added to the argument; in her book 'Dark Matters: Stories on the Surveillance of Black People', Simone Browne (2015), Professor of Black studies at the University of Texas at Austin, illuminates the conditions of Blackness as a key locus through which surveillance is practiced and resisted. Drawing on Black feminist theories, sociology and cultural studies, Browne's book provides a compelling illustration of how the history of racial formation, and the violence implicit in it, is reflected into the contemporary practices and technologies of surveillance. Identification practices, which emerges from the book, are themselves part and parcel of racialised surveillance, and are to be studied in terms of the affordances of racial discrimination that they yield.

It is primarily from the works of Browne (2015) and Costanza-Chock (2020) that my reaction to the idea of a 'dark side' of IT has emerged. Both works are clear on how, if technology shapes core aspects of people's lives, its harmful effects cannot be conceived as incidental; they are written directly in how the artefact is designed, and design is functional to strengthening existing forms of marginalisation. The digital ID systems discussed in this book, with their effects on fundamental rights, underscore the relevance of a human rights perspective; assessing digital ID systems in terms of human rights impact is crucial to make sense of the harm and discrimination that the same systems can result in. This makes Human Rights Impact Assessments (HRIAs), a type of assessment grounded in international human rights law, a tool of strong relevance for digital ID design and evaluation.

The World Bank and Nordic Trust Fund (2013) define a HRIA as 'an instrument for examining policies, legislation, programs and projects to identify and measure their effects on human rights'. The purpose of a HRIA is to use a systematic process to investigate 'the potential impact of laws, policies, programs, projects, and interventions on the human rights of individuals, population groups, or general society' (McCall-Smith, 2022, cited in Haki na Sheria, 2023: 6). HRIAs have long been used in the field of development studies; as noted by the World Bank and Nordic Trust Fund (2013), they have been an essential route to making human rights considerations operational in many legal and policy contexts. This tool reflects a growing effort by the human rights community to operationalise the relevance of human rights in various fields, seeking to generate a structured understanding of the direct and unintended consequences of policies and other interventions on human rights (World Bank and Nordic Trust Fund, 2013).

Importantly, the most distinctive aspect of a HRIA is that of being grounded in binding principles of international human rights law. Other forms of impact assessment exist in development programmes; for instance, Environmental Impact Assessments (EIAs) and Social Impact Assessments (SIAs) are conducted across countries to evaluate multiple interventions, and they are implicitly underpinned by human rights values (World Bank and Nordic Trust Fund, 2013: 11). But, notes the same report, the normative framework of a HRIA marks the difference from other approaches; building assessment on human rights law adds legitimacy and accountability to the whole exercise, leveraging human rights as the dominant language for social justice claims in many parts of the world. A HRIA is therefore positioned to provide uniquely significant recommendations, and to limit the acceptability of trade-offs on recommendations made (World Bank and Nordic Trust Fund, 2013: 11).

Beyond the normative human rights framework, the World Bank and Nordic Trust Fund (2013) list five more essential elements of a HRIA. A second element, public participation, seeks to ensure the active participation of stakeholders throughout the phases of an intervention, from design to evaluation. A third element, equality and non-discrimination, points to the need to assess interventions for their ability to grant equality and produce non-discriminatory results on users. A fourth, related element is on transparency and access to information; both the contents of the intended project, and those of the criteria used in the HRIA, should be transparently accessible to intended stakeholders, in order to truthfully understand the effects the intervention may yield. Two final elements are also interrelated: accountability, meaning the establishment of effective mechanisms that enable redress in case human rights are undermined, and an intersectoral approach, which sees all rights – civil, political, economic, social and cultural – as interdependent. Such coexisting elements, found across many HRIAs (World Bank and Nordic Trust Fund, 2013: 12), maximise their potential as a tool in the armoury of human rights protection.

While largely utilised in the fields of policy and development studies, few HRIAs have been conducted in relation to digital ID programmes. In some cases, implementation principles have differed markedly from those of a HRIA, reflecting instead a logic of fast and quiet development of digital ID systems. The Centre of Human Rights and Global Justice reports on the words of Aadhaar's founder, Nandan Nilekani, in relation to the inception of the project:

> Our view was that there's bound to be opposition, right, so that's a given. So we said how do we address that? One was, do it quickly. Because if you do it quickly it's less likely to coalesce against you. The second was do it quietly, get it done. And third was we said that in any case there is going to be a coalition of opponents, so is there a way to create a positive coalition of people who have a stake in its success? (Centre of Human Rights and Global Justice, 2022: 84).

At the same time, a logic of 'do it quickly' and 'do it quietly' seems problematic in a HRIA perspective. In unpacking this, Berkeley's International Human Rights Law Clinic and Haki na Sheria (2023) provided one of the first instantiations of a HRIA on a digital identity scheme; in the wake of the devising of UPI as a novel digital ID scheme in the nation, the organisations identified HRIAs as a possible tool to better evaluate the human rights risks of Kenya's proposed digital ID. The problem, reflects the report, is first of all in terms of the potential discriminatory effects that UPI may yield: double registered persons, 'border' and Muslim communities are especially at risk, undergoing the invasive 'vetting' processes described in Chapter 7. With the risk of exacerbation of long-term discriminations, a digital ID system in Kenya would benefit from a HRIA, whose principles of transparency and participation contrast the imperative of acting 'quickly' and 'quietly' (Centre of Human Rights and Global Justice, 2022: 84–85).

Two more aspects of the same report illuminate the potential of HRIAs in the digital identity space. A first aspect goes back to the World Bank and Nordic Trust Fund (2013); being grounded in binding international human rights law, HRIAs derive legitimacy and accountability from the same law, thereby drastically limiting the acceptability of trade-offs. In the case of Kenya, a HRIA may complement the Data Protection Impact Assessment required by the High Court in the 2020, in relation to the challenge and subsequent demise of the Huduma Namba scheme (Haki na Sheria, 2023). The presence of long-standing discriminatory dynamics, combined with the risks of exacerbation of the same dynamics through a new digital ID scheme, make Kenya's UPI a strong candidate for a HRIA, which if conducted could involve stakeholders in the direct assessment of the programme's risks and benefits.

Secondly, the report notes that international law imposes a requirement on Kenya to conduct a HRIA. Kenya is required to evaluate the human rights impacts of government programmes (Haki na Sheria, 2023: 11). This includes UPI, on which the Government is called to address any potential human rights risks. Kenya's human rights obligations are reinforced by pursuance to treaties such as International Covenant on Economic, Social and Cultural Rights (ICESCR) and the African Charter on Human and People's Rights (ACHPR), reinforcing the requirement for a HRIA (Haki na Sheria, 2023: 11). The reports' recommendations turn into practice the basic elements of a HRIA; requirements for access, accountability and public participation lead the two organisations to recommend pausing ongoing digital ID projects, till the human rights impact of UPI is publicly, transparently and comprehensively assessed (Haki na Sheria, 2023: 18).

The grounding of HRIAs in international human rights law, along with ability to map risks connected to interventions of various types, positions HRIAs as a first potential route to *reversing dark matter* in digital ID. On the one hand, HRIAs offer an analytical instrument which directly faces the risk of harm, devising precise routes to appraising it before a potentially dangerous intervention is launched. On the other, the same instrument affords imagining ways to avoid such harm; these can mean pausing risky programmes, but also devising interventions that, mindful of the risks, propose actionable paths to avoiding them. This makes HRIAs a relevant tool in the imagination of fair ID.

Reversal 2: Anti-injustice ID Artefacts

The making of HRIAs has illuminated one possible route to fair ID. It consists of assessing the human rights risks that a digital identity system may pose, thereby enabling designers and decision-makers to caution against such risks. But more routes to fairness, rather than risk assessment, may involve direct action on the system that produces injustice. Returning to the politics of anti-poverty artefacts as in Winner (1980) leads to a key point; building fairness does not necessarily mean rebuilding the artefact of identification, but acting on one or more components of the broader system of which it is a part.

Two stories, coming respectively from the Indian states of Chhattisgarh and Tamil Nadu, illustrate this point. These are stories where just features were designed in the ecosystem around digital ID, where an *ecosystem* refers to (a) all components surrounding the digital ID architecture and (b) all the steps that the person traverses within a digital ID system. As noted with the politics of anti-poverty artefacts, digital ID will not alter existing policies; on the other hand, it may crystallise existing injustices, such as the exclusion errors discussed in Chapter 4. It is hence relevant to study fairness across the different parts of ID ecosystems, and of the programmes they enable; for instance, the determination of goods received under social protection schemes and the way the same schemes are structured.

Research conducted by Raghav Puri sheds light on this point. In December 2011, Puri published on the Indian magazine Frontline an article titled *Loud No to Cash*: the article was written when citizens of Chhattisgarh were asked about their preference between the PDS and the cash transfers that promised to infuse fairness in the food subsidy system. Chhattisgarh did not historically have a strong PDS; in fact, for many years the state ranked bottom of the national scorings for PDS offtake (Drèze & Khera, 2015). As my research also highlighted in other states, the poor quality of food grains in some villages meant people steered away from the PDS, which was a strong disincentive to consider the programme as a viable source of nutrition.

But studying Chhattisgarh, Puri found PDS users to be strongly averse to a switch to cash transfers. This could be surprising: a poor PDS record, with poor quality and offtake over the years, in principle gave little reason to defend the system. But Chhattisgarh was one of the most actively reforming states identified by Drèze and Khera (2015). PDS reforms at the state level have significantly strengthened the food distribution system, making its reception highly positive among users. First, under the government of Raman Singh in 2004, the state introduced the Chhattisgarh PDS control order, which shifted the management of ration shops from private dealers to community organisations. This generated greater accountability, because those running the shops were people from the villages (Puri, 2012: 21). Also, putting ration shops in the hands of public entities like cooperatives, gram panchayats (village councils) and women's self-help groups delinked the ration shop from a profit-making logic, which reduced the incentive to diversion that previously motivated high leakage rates.

The second reform embarked upon in Chhattisgarh pertained to the transport of goods from the government godowns to PDS outlets, a process that, notes again Puri (2012: 21), initially happened through private trucks. A substantial transparency problem was posed by lack of clarity on where commodities allocated from the central government were being disbursed to, which left wide room for diversion from the PDS. The state found, however, one solution to the problem, using trucks that delivered food grains directly to ration shops; differently from trucks owned by private companies, these were painted yellow, to visibly show where trucks aimed at ration shops were really headed. Doorstep delivery to ration shops, enabled by yellow-painted trucks, dismantles the idea that controls should take place at the last mile alone; while systems like ABBA act at the ration shop level, doorstep delivery addresses the problem at its early stages. With this move, 88% of the respondents surveyed by Puri (2012: 21) reported getting their food grains regularly at the correct prices.

But a third reform also took place under Raman Singh. After the national move to a targeted PDS in 1997, the 2002 below-poverty-line (BPL) survey excluded many households from being classified as BPL. This was because the Planning Commission put caps on poverty rates, meaning only a limited number of households could be classified as BPL. In response the government of Chhattisgarh introduced the

Mukhyanmantri Khadiyann Sahayata Yojana (MKSY), which translates as the Chief Minister food relief scheme, in April 2007. Under the scheme, the government provided ration cards to additional 1.9 million households, which were recognised as BPL in the 1991 and 1997 surveys, but capped out in 2002 (Puri, 2012: 23). This brought the Chhattisgarh PDS, even before the promulgation of the NFSA in 2013, a lot closer to universality than it previously was; it made over 80% of the rural poor entitled to the PDS, supplanting the fallacies of the BPL capping by the Planning Commission. The result, concludes Puri (2012: 21), is a system where people conveyed, in the PDS Survey 2011, an overwhelming 93% preference for the PDS over cash transfers, way before the introduction of biometric recognition across states.

Insights from Chhattisgarh induce further thinking on the design of the PDS. The reforms made under Raman Singh move in the opposite direction from targeting. On the one hand, targeting embodies the core principle of digital identity, by subordinating people's access to systems to their identification. On the other, why is it that a programme aimed at a universal right, such as food, has been targeted to categories? This poses the fundamental question on the ethicality of targeting (Devereux, 2016), and in turn on the authorisation-authentication nexus at the basis of digital ID. Can digital ID, on the other hand, support the enforcement of universal food security systems?

The story of Tamil Nadu, the only Indian state that preserved a universal PDS, speaks to the effects of this choice. In this state, different cards are provided according to the status of households under the NFSA: users with different poverty status have different entitlements, while a basic entitlement is guaranteed to everyone. A fundamental difference is between a so-called *rice card* (all-commodity card) and a *sugar card* – all-commodity cards (green) are issued to those cardholders opting for rice as well as all other essential commodities, while sugar cards (white) are issued to cardholders opting for sugar instead of rice. Sugar cardholders can buy all other essential commodities except rice, and also get three more kg of sugar in lieu of rice. As the state shows, a layering of need levels can be achieved even in a universal context, and it is this layering that can be crystallised through the smart card system studied by Carswell and De Neve (2022).

The case of Tamil Nadu illuminates a fundamental aspect in imagining fair ID. When connected to a universal system, ID becomes a legitimiser of rights, rather than a mechanism subordinating them to people's digital identification and authentication. In the Tamil Nadu case, fair ID reflects the fair politics characterised by expansion of rights to food access. Smart cards become, in this space, a legitimiser of a universal right; fairness is found in artefacts that, even as analogue as the yellow-painted trucks of Chhattisgarh, take direct measures for transparency towards users.

Fair ID: A Conceptual Apparatus

So far we have seen two ways of imagining fair ID. One relies on HRIAs as a route to protection from the human rights risks associated to digital ID systems, building protective measures grounded in international law. Another one requires acting on the ecosystem of which digital ID is an integral part, in ways that affect the politics coded in the artefact. Both ideas are premised on the goal of protecting people from ID-induced harm, and inspire the construction of fair ID.

Both routes are, however, indirect ways of dealing with ID unfairness. One is designed to protect users from human rights violations and the other affects the broader system in which the technology is placed; none of them, on the other hand, suggests direct ways to design digital ID in the light of fairness. There have been attempts to doing so, for instance in the case of smart cards in Tamil Nadu; designed with an embedded code, they sought to remove the issues of ABBA's fingerprint-based recognition. But as noted in this chapter, smart cards reproduced the issue of informational injustice, replacing physical delivery stamps with a text message that may, or may not, be received by the user. An artefact aimed at combating one type of injustice resulted, note Carswell and De Neve (2022), in other forms of it, ultimately not supporting the thesis of a fairer artefact in the eyes of users.

This leaves us with limited evidence of artefacts that have been thought out, tested and redesigned to contrast the injustices studied in this book. But at the same time, recent research is substantiated of important concepts to study fair ID. These are concepts that speak about dignity, digital rights and that connect the discourse of digital ID to that on algorithmic fairness and artificial intelligence (AI). The coda of this chapter engages these concepts, relying on emerging research in which, I argue, we find the key components of a conceptual apparatus for studying fair ID.

Data for Dignity: Reimagining Digital Social Protection

The notion of *data for dignity*, as put forward by Joan López (2022), makes an important contribution to the discussion on fair ID. As noted in Chapter 5, in April 2021 the Colombian Government introduced Sisbén IV, a new version of the system (Sisbén) that assigns households to different categories based on prosperity scoring. The system classifies households based on self-reported data collected in surveys, and its classification is used to select the beneficiaries of 19 government social programmes; these cover schemes in healthcare, poverty alleviation and housing, and make the Sisbén classification a critical determinant of the welfare that a household is entitled to receive.

Shortly after the introduction of Sisbén IV, however, numerous households raised the issue that their classification had increased with respect to previous versions. Increases in the classification resulted in households being denied essential social services; when checking into the system, many users found their status as pending 'information verification', and were barred from social welfare provisions for this reason. While this brings back the informational injustice problem studied in Chapter 5, the issue also lies in the response that the government, across multiple social media accounts, gave to such complaints. To many claims, the government gave one overarching answer 'any claims presented by individuals should be regarding the information recorded, but not the classification' (López, 2022: 8).

At least two aspects of the problem speak to injustices of digital ID. First, the system shifts the locus of decision-making to a form of algorithmic power that is invisible to the user; the classification, implies the logic of Sisbén IV, is correct, because the system is built to convert data into the right policy decisions behind it. Seen with Winter (1980), the politics of the artefact has an impositive logic, which denies people access to the survey and even stigmatises respondents as liars, as it happened to people accused of defrauding the system (López, 2022: 8). This deprives households of the ability to challenge decisions made on them, due to the informational injustice built in the artefact, but it also deprives households of dignity, intended as the human quality of being worth of respect. Human dignity, notes López (2022), is denied to people claimed to accept the decisions of an algorithm which is invisible, but which may result in denial of the most basic social assistance.

A second aspect lies in the algorithm that Sisbén IV, unlike previous versions of Sisbén, is built on. The point is made in earlier research by López (2021b), programmatically titled 'Experimentando con la pobreza' (experimenting with poverty). As illustrated there, Sisbén IV shifts from a system that verified the living conditions of households to an algorithm that, based on survey data, *predicts* their income probabilities. To do so, Sisbén IV uses another classification, the Great Integrated Household Survey conducted by the National Administrative Department of Statistics (DANE) to build profiles of the socioeconomic features of households, and of their potential income (López, 2022: 23). The reason behind the change, notes López (2022), is one of trust – the system reflects trust in the data reported to DANE rather than those reported to Sisbén. The qualification of Sisbén in terms of the profiling and punitive logics of social welfare algorithms, theorised in Eubanks (2018), adds to the de-dignification of poor households, deprived of the ability to challenge decisions that affects their access to services.

But the system, López (2022) continues, has in itself the basis of change. As noted in the discussion of HRIAs, respect for human rights is upheld in international law, and should be reflected in all phases of the life cycle of policy design. Datafied systems, it has been noted, can reflect logics that deprive people of dignity; but they can also be plied to restoring justice, by building mechanisms for users to be actively included in the decisions made about them. The idea of 'data for dignity' can be directly embedded in algorithms; to do so, concludes López, requirements of transparency, accountability and participation should be inbuilt in the system, making the process visible to users and combating the informational injustice associated to it.

It is in these requirements that the idea of 'data for dignity' is directly substantiated. The idea speaks to Cheesman's (2022a) notion of infrastructure justice; this pointed to the uneven benefits that sociotechnical systems entail, unevenness that Sisbén beneficiaries, denied information on the basis for their classification, directly experienced. But at the same time, it positions the system as a locus where dignity, based on the principles of international human rights law, can be restored. It is human dignity that datafied systems can contribute to rebuild, and a fair ID, giving the person the dignity of participation in key decisions made on them, can be imagined in this light.

Digital ID and Algorithmic Fairness

López poses a problem of algorithmic fairness, showing how closely connected this is to digital ID. His work invites engagement with the notion of AI, defined by Floridi (2023) as a growing resource of interactive, autonomous and often self-learning agency. The case of Sisbén IV illustrates how the separation of digital ID from research on welfare algorithms is not anymore possible; on the one hand, bodies of work on data justice and digital ID have so far followed separate paths as noted in Chapter 3. But as core decisions affecting people's entitlements, such as the scoring of households in Sisbén IV, are shifting to algorithmic power, digital ID research needs to rely closely on what research on AI for welfare has to share with it. Two examples centred, respectively, on India's PDS and a cash transfer programme in Jordan substantiate this point.

India's PDS has itself born the consequences of algorithmic evolution. In January 2024, Tapasya, Kumar Sambhav Srivastava and Divij Joshi published a study of Samagra Vedika, an algorithmic system used by the state of Telangana to cut out illegitimate beneficiaries of the PDS and identify legitimate ones. Behind the algorithm is a simple rationale; biometric verification through Aadhaar, as noted in Chapter 4, is meant to help weeding out non-entitled users. But Aadhaar-based authentication does not ensure that claimants are entitled, and needs to be complemented by forms of verification such as the Aadhaar seeding of ration cards. To enhance verification and build the initial lists of eligible beneficiaries, several Indian states have turned to welfare algorithms. Similar to Sisbén IV, they cross people's data across multiple databases, finally arriving at decisions on whether subjects are entitled to particular welfare provisions.

This is the case for Telangana's Samagra Vedika. Similarly to other forms of digital identification, the algorithm was originally not conceived for welfare purposes. Its birth is traced to 2016, when it was built for the Hyderabad Police to identify people of interest for criminal verification. Using algorithms developed by the private company Posidex, it was introduced in the PDS in 2016 as a pilot, and in 2018 it was adopted for most of the state's welfare schemes (Tapasya et al., 2024). The algorithm, sustains the Telangana government, has led to substantial savings from identified leakages. Such praise is echoed by the Department of Information Technology in the state, which notes its cruciality to a fair distribution of entitlements (Tapasya et al., 2024). Arguments raised on early technologies applied to the PDS, culminating into the incorporation of Aadhaar in it, are reflected in the same statement, invoking the logic of fight against leakage for the introduction of AI tools in a long-standing food security programme.

The investigation by Tapasya et al. (2024), however, casts doubt on such statements. Their research illuminates multiple cases of unfairly excluded people; such as Bismillah Bee, a 67-year-old widow whose deceased husband, a rickshaw driver, was erroneously tagged as a car owner, resulting in the family losing PDS entitlement. The man, further research uncovered, had been mistaken for a car owner with a name similar to his. In addressing Bee's case, however, the authorities went with the decision that the algorithm made. As Tapasya et al. (2024) continue to note, Bee's case is epitomic of many more: from 2014 to 2019, the state of Telangana cancelled over 1.86 million existing ration cards and rejected 142,086 new applications without notice. While the government claimed the technology to be of 'high precision', a Supreme Court-imposed re-verification of cancelled cards in April 2022 suggested that at least 7.5% of the cards were wrongfully rejected (Tapasya et al., 2024), which reminds of the legal injustice of digitally induced exclusions seen in Chapter 4.

Another form of injustice is echoed by the same study. López (2022) posed a problem of algorithmic opacity. In the case of Samagra Vedika, the source code of the algorithm remains private, even after a Right to Information request to make it open (Tapasya et al., 2024). Motivations for this are related to Posidex's ownership of the algorithm, which, it is stated, does not offer the legal basis to make it public (Tapasya et al., 2024). This results in another case of informational injustice; people do not know how Samagra Vedika crosses their data, and have no effective means to contest the decision made by the algorithm. Power imbalances, noted earlier for PDS users compelled to Aadhaar enrolment, are reproduced in the light of AI; key decisions are charged to the algorithm, leaving people with limited power to oppose or meaningfully question them.

A recent case of automated cash transfers in Jordan reflects similar concerns. In June 2023, Human Rights Watch released a report titled '"Automated Neglect": How the World Bank's Push to Allocate Cash Assistance Using Algorithms Threatens Rights', which documents the human rights impact of a World Bank-financed automatised cash transfer scheme. Known as the United Cash Transfer Program, but commonly referred to with its original name 'Takaful', the scheme uses an algorithm that screens out families that do not meet the basic eligibility criteria. It then identifies which of those remaining should receive cash transfers, by ranking their level of economic vulnerability. In input, the algorithm takes data from 37 government agencies to create economic profiles of families, in output it ranks them from less poor to poorest, selecting a limited number of households, classed as 'most vulnerable', as cash recipients.

But the algorithm, notes Human Rights Watch, results into invisibilisation of the complexities behind answers given in the questionnaires. Several interviewees note, among the main concerns, that they can only declare an income that matches their living expenses. The National Aid Fund, in charge of the scheme, notes that it requires to declare living expenses not exceeding 20% of income. Gaps between expenses and incomes, bridged for example by family support, are not reflected in calculus; this affects both targeting accuracy, and people's reporting behaviour in order to preserve their eligibility (Human Rights Watch, 2023). Similar to Samagra Vedika, the algorithm also flags the value of assets including vehicles, livestock or businesses to weed out ineligible users. Such a function results, even in this case, in exclusions ratified by algorithmic decisions; this imposes a choice between right to social security and the achievement of rights to 'a decent living, health, and food' (Human Rights Watch, 2023: 43).

The connection of digital ID with welfare algorithms invites two more reflections. A first one pertains to the novel positioning of digital ID in AI research, noting that much older discourses, such as the use of technology in the fight against leakage, are being replicated in the AI space. A second one concerns the link of digital ID to the discourse of algorithmic fairness that the AI world has engaged (Marabelli, 2024). On the one hand, digital identification and automated targeting have operated since long before AI. On the other, stories from Sisbén, Samagra Vedika and Takaful show the profound connection of fairness in the ID space with fairness of the algorithms in which such IDs participate, algorithms to which new forms of resistance are developing over time (Bonini & Treré, 2024). This is why digital rights make a fundamental part of the discourse on fair ID.

Fair ID in the Digital Rights Space

Taylor (2017) notes how, just like an idea of justice is needed to establish the rule of law, an idea of data justice serves to 'determine ethical paths' in a digital world. In a similar logic, a digital world requires the establishment and upholding of *digital rights*, a concept that Roberts et al. (2021) define in terms of human rights in online spaces. In their study of online civic spaces in ten African countries, Roberts et al. (2021: 11) note multiple manifestations that digital rights, and their denial, can acquire; these pertain to 'the right to privacy, freedom of opinion and speech, freedom of information and communication, gender rights, and the right to freedom from violence'. As they convert human identities into digital data, digital identity systems are an integral part of a digital rights discourse, within which emerging narratives of fair ID are positioned.

The discourse on digital rights is closely associated to Internet access, and positions Internet shutdowns as a violation of a multiplicity of freedoms, as well as human rights (Access Now, 2023). The same relation has been elaborated on by Fabio Cristiano, Assistant Professor of conflict studies at Utrecht university. Through his studies on cyberspace in Palestine, he illustrates the digital mediation of multiple aspects of political life and liberties; such mediation is exacerbated in the Palestine–Israel conflict, where Israel's military occupation involves the power to 'kill-switch' the Internet in the occupied territories (Cristiano, 2019: 257; 2022). The ongoing military assault of Israel on Gaza, started in October 2023, involves extensive use of such power, with Internet access in the Gaza strip being switched off for extensive stretches of time (PalTel Group, 2023) and discontinuously enabled and disabled, undermining any ability to effectively rely on the presence of connectivity. The case of Palestine is epitomic of the complexity of digital rights landscapes, and as it concerns the object of this book, of the way digital ID is positioned within them.

Recent work from the civil society space offers important depictions of such a landscape. The report 'Mapping Humanitarian Tech: exposing protection gaps in digital transformation programmes', published by Access Now in February 2024, explores it from the angle of partnerships engaged by humanitarian actors with private corporations, of which the WFP–Palantir case, discussed in Chapter 6, is an instantiation. Written by senior humanitarian officer Giulio Coppi, the report illustrates how such partnerships affect the digital rights of the affected communities; along with sensitive data management, connectivity and cybersecurity, digital ID is a core area of its enquiry. The investigation is placed in the context of the increasingly transactional dynamic of the relation of humanitarian actors with recipient communities. While private actors play a substantial role in such a relationship (Martin et al., 2023) limited knowledge is available on the actors in point, and on their patterns of engagement with humanitarianism (cf. The Engine Room, 2023). This limits, notes the report, the ability of civil society to address crucial aspects in the aid–tech relation (Access Now, 2024).

Crucially for this book, the report underscores important dimensions of the belongingness of digital ID to the broader discourse on digital rights. The WFP–Palantir partnership case is again epitomic of this; humanitarian data collection systems are increasingly based on the digital identification of recipients, to which the provision of essential services is made conditional (Madon & Schoemaker, 2021). This is how humanitarian agencies, interfacing with privates, built extremely vast digital ID databases; WFP's Scope hosted data from more than 63 million vulnerable people in 2020, and only in 2022, UNHCR reported more than 3 million individual registrations by its implementing partners through the PRIMES proGres v4 system (Access Now, 2024: 31). Participation of private entities to the humanitarian aid equation arguably results in the *sector transgressions* that Taylor et al. (2023) refer to; transgressions that, in the name of a logic of increased effectiveness, provide private actors with global insights on vulnerability. Questions on use of the same insights towards market decisions, against the backdrop of the aidwashing logic studied by Martin (2023), remain open.

In addition, Access Now (2024) raises the point on how access to digital and communication systems has become increasingly conditional to owning a digital form of ID. For some countries, the registration of a SIM card has become conditional to proof of residency: as noted by Martin and Taylor (2021a), the same logics has made it illicit to sell SIM cards to Rohingya refugees in Bangladesh, resulting into active limitations of digital connectivity for vulnerable populations. In 2018, Egypt made the purchase of a SIM conditional to showing a valid passport or residency permit; joint action by aid actors and private companies subsequently enabled recognition of UNHCR refugee credentials as a valid proof. But the exception, notes again Access Now (2024: 32), was a 'corporate deal' conditional to UNHCR acting as trusted partner, providing names and phone numbers of beneficiaries. Such a conditionality reiterates the issues of informational injustice raised in Chapter 5; users are not positioned to meaningfully question use of their data, but find themselves in the position of having core services denied if data are not provided. Within the imagination of Fair ID, questions of data ownership and consequences of sharing of personal information across providers of the private sector remain to be answered.

Summary

Learning from the stories of injustice and resistance narrated in the book, this chapter has put forward a conceptual apparatus for imagining fairer forms of digital ID. Cheesman's (2022a) notion of *infrastructure justice* is central to this apparatus. Emerging research on data for dignity, algorithmic fairness and digital rights has contributed to its construction, illuminating important aspects of how fair ID can be enacted. This book has been, since the beginning, a hymn to hope; the hope that by understanding unfairness in digital ID, and the paths of its effects on people, a fairer world can be built for the digitally identified. The conceptual apparatus suggested in this chapter contributes to imagining such a world.

CONCLUSION: INTERLOCKED LENSES

When I started working on this book, I had wished its conclusions would show that overcoming unfair ID could strengthen fundamental rights where these are threatened. The state of the world does not support this conclusion. Denial of digital rights is generating diverse forms of oppression, which proliferate globally; the ongoing war on Palestine, in which the Gaza Strip is subject to a long telecommunications blackout, dramatically illustrates the problem. The coupling of digital and human rights has become increasingly tight. As a result, the denial of digital rights intersects with attacks to human rights, of which the situation of Palestine is epitomic. The Internet shutdowns that started in October 2023 go in parallel with heavy bombing and the suspension of food, fuel, water and medical aid, with over 28,000 civilians killed since the beginning of Israeli bombings in October and many more injured, missing or in extremely precarious life conditions (Amnesty International, 2024).

It can be argued that the current international situation is metonymic of a wider problem. With the violence that characterises it, the Palestinian case is the latest in a long string of attacks on Internet freedom. In its latest yearly report on the matter, Access Now (2023) recorded 187 Internet shutdowns in 35 countries, with 48 shutdowns coinciding with human rights abuses in 14 countries in 2022 alone. With its welfare and development promises, digital ID has been the conceptual object of this book and at the same time, it cannot be seen in isolation from the digital rights of the people it affects. Internet shutdowns ontologically show this: their denial intersects with negation of the most fundamental human rights, the same that instantiations of unfair ID have illustrated. The production of injustices through unfair ID epitomises a much larger phenomenon, in which violations of rights in the physical space are transposed to the digital, and further reified through IT artefacts.

In this landscape of rights' violation and fight for redressal, the data justice lens has been a guiding light to make sense of complex digital ID phenomena. It was the data justice perspective that equipped us with the lexicon used in this book, and that helped us make sense of a literature – that on digital ID – whose sparsity across fields can confuse the reader. Different thematic foci in the digital ID domain tend, in addition, to create islands of knowledge that do not speak to each other. This jeopardises the domain's ability to see the impact of digital ID on people and of voicing the real perspectives lived by its users. In a world in which digital ID research is sometimes perceived, as I heard at a conference, as 'following the money', my lens has taken the opposite perspective; a data justice view has allowed me to interrogate what happens to the user, at a time when it is on them that multiple harmful effects are produced through the artefacts of identification. It is a users' perspective that a data justice lens has illuminated, and that the framework proposed here has sought to advance.

In finishing this book, some considerations need to be made on the concepts it has relied on. The first pertains to the fundamental role played by early-career researchers in presenting the picture of digital ID that the book has given. I come from a scientific domain that makes this point provocative; in my field, Information Systems, 'standing on the shoulders of giants' is a prerogative point that is hardly ever questioned. But this book has, on the other hand, shown that concepts for understanding ID unfairness come from a pool of international, early-career colleagues, that few fora – with a special mention for the Surveillance Studies Network and Data Justice conferences, organised respectively by the international Surveillance Studies Network and the Data Justice Lab at Cardiff University – gather together. Seminal research on digital ID, and on the socio-technical foundations of this concept across fields, remains influential; but what the book has advanced is the value of novel research and of how it grapples with the problems of digital ID unfairness articulated through these pages. In the complexity of the digital ID apparatus, it is in early-career colleagues that I found an anchor, and in their conceptualisations that I encountered the guiding light of the debates presented here.

A second consideration is in that the two lenses twinned in this book, those of *data justice* and *data activism*, are inseparable from each other in the narration of unfair ID. It took the whole work through the manuscript to understand their twinning: data justice and data activism are facets of the same coin, which in the absence of each other provide an incomplete narration of the story. Through data justice, we unpacked the unfair manifestations of digital ID towards people; but through data activism we illuminated resistance, and the shapes it takes when placing ID in an increasingly datafied world. Dimensions of data activism are in turn interlocking, with the indivisible combination of *proactive* and *reactive* components (Milan & Van der Velden, 2016); the reactive part of data activism adopts diverse means to opposing digital ID injustice. But the proactive part lies in imagining fair ID, building anti-injustice artefacts that devise, in their very features, paths to combating ID unfairness.

I stated, from the beginning of this manuscript, that this book is a hymn to hope. I only wanted the book to end at a time of hope, rather than one in which denial of digital rights is coupled with one of the most violent genocides that history has ever witnessed. In a world where digital tools are plied to violence, and on its blind perpetration over civilians, the power of anti-injustice artefacts provides a light of hope in building a road to freedom. It is a light through the darkness that the harmful outcomes described in this book have detailed; combating unfair ID means taking stock of injustice, but also of the hopeful light of resistance that communities across the world have enacted. It is with resistance that the journey continues. Only through it can a fair ID be made.

REFERENCES

Abuya, K. (2023). Kenya discontinues Huduma Namba, takes another try at digital identities. https://techcabal.com/2023/06/02/kenya-to-launch-new-digital-id/

Access Now (2020). *National Digital Identity Programs: What's Next?* https://www.accessnow.org/wp-content/uploads/2019/11/Digital-Identity-Paper-Nov-2019.pdf

Access Now (2021). *Busting the Dangerous Myths of Big ID Programs: Cautionary Lessons from India.* https://www.accessnow.org/wp-content/uploads/2021/10/BigID-Mythbuster.pdf

Access Now (2023). *Weapons of Control, Shields of Impunity: Internet Shutdowns in 2022.* https://www.accessnow.org/wp-content/uploads/2023/05/2022-KIO-Report-final.pdf

Access Now (2024). *Mapping Humanitarian Tech: Exposing Protection Gaps in Digital Transformation Programmes.* https://www.accessnow.org/wp-content/uploads/2024/02/Mapping-humanitarian-tech-February-2024.pdf

Access Now and 18 other signatories (2022). *Open Letter to the ITFlows Consortium: Stop Tech Tools for Predicting Migration that can Be Repurposed to Violate Fundamental Rights.* https://www.accessnow.org/press-release/open-letter-itflows-consortium/

Access Now and 30 other signatories (2021). *Fundamental Rights Concerns about the EURODAC Reform. Letter to Mr Buxadé, MEP, Shadow Rapporteurs, and Members of the Committee on Civil Liberties, Justice and Home Affairs (LIBE)*, 8 September. https://www.amnesty.eu/wp-content/uploads/2021/09/EURODAC-Open-letter-8-Sept-2021.pdf

Access Now and 45 other signatories (2022). *Open Letter: World Bank and Its Donors Must Protect Human Rights in Digital ID Systems.* https://www.accessnow.org/press-release/open-letter-to-the-world-bank-digital-id-systems

Access Now and eight other signatories (2023). *Past Learnings Must Be 'at the Heart of Implementing' a Digital Identity System in Kenya.* https://www.accessnow.org/press-release/kenya-digital-identity-systems

Access Now (n.d.). *Digital ID Systems: The Global #WhyID Campaign.* https://www.accessnow.org/campaign/whyid

Aggarwal, A. (2011). The PDS in rural Orissa: Against the grain? *Economic and Political Weekly*, 46(36), 21–23.

Ahluwalia, D. (1993). Public distribution of food in India: Coverage, targeting and leakages. *Food Policy*, 18(1), 33–54.

Akbari, A. (2021). Authoritarian surveillance: A Corona Test. *Surveillance and Society*, 19(1), 98–103.

Aker, J. C., Boumnijel, R., McClelland, A., & Tierney, N. (2016). Payment mechanisms and antipoverty programs: Evidence from a mobile money cash transfer experiment in Niger. *Economic Development and Cultural Change*, 65(1), 1–37.

Allu, R., Deo, S., & Devalkar, S. (2019). Alternatives to Aadhaar-based biometrics in the Public Distribution System. *Economic and Political Weekly*, 54(12), 30–37.

Amnesty International (2020). *Failing to Do Right: The Urgent Need for Palantir to Respect Human Rights*. https://www.amnesty.org/en/documents/amr51/3124/2020/en

Amnesty International (2024). *Israel/OPT: New Evidence of Unlawful Israeli Attacks in Gaza Causing Mass Civilian Casualties amid Real Risk of Genocide*. https://www.amnesty.org/en/latest/news/2024/02/israel-opt-new-evidence-of-unlawful-israeli-attacks-in-gaza-causing-mass-civilian-casualties-amid-real-risk-of-genocide/

Appadurai, A. (1986). Introduction: Commodities and the politics of value. In Appadurai, A. (Ed.), *The Social Life of Things. Commodities in Cultural Perspective* (pp. 3–63). Cambridge: Cambridge University Press.

Article 14 (2022). *Millions of Children Will Soon Need Aadhaar IDs to Access Their Right to a Nutritious Meal*. https://article-14.com/post/millions-of-children-will-soon-need-aadhaar-ids-to-access-their-right-to-a-nutritious-meal–62bc915131cc9

Baez, J. E., & Camacho, A. (2011). *Assessing the Long-Term Effects of Conditional Cash Transfers on Human Capital: Evidence from Colombia*. World Bank Policy Research Working Paper, 5681. SSRN-id1865119.pdf

Bardhan, P. (2011). Challenges for a minimum social democracy in India. *Economic and Political Weekly*, 46(10), 39–43.

Baxi, P. (2019). Technologies of disintermediation in a mediated state: Civil society organisations and India's Aadhaar project. *South Asia: Journal of South Asian Studies*, 42(3), 554–571.

Bayly, C. A. (2000). *Empire and Information: Intelligence Gathering and Social Communication in India, 1780–1870*. Cambridge: Cambridge University Press.

Bennett, C. J., & Lyon, D. (2013). *Playing the Identity Card: Surveillance, Security and Identification in Global Perspective*. London: Routledge.

Better Than Cash Alliance (2021). *Colombia's Ingreso Solidario: Public-Private Collaboration in Covid-19 Emergency Payments Response*. https://www.betterthancash.org/news/learning-series-covid-colombia

Bonini, T., & Treré, E. (2024). *Algorithms of Resistance: The Everyday Fight against Platform Power*. New York: MIT Press.

Border Violence Monitoring Network (2024). *Decoding Balkandac: Navigating the EU's Biometric Blueprint*. https://borderviolence.eu/reports/balkandac/

Breckenridge, K. (2014). *Biometric State: The Global Politics of Identification and Surveillance in South Africa, 1850 to the Present*. London: Cambridge University Press.

Breckenridge, K. (2019). Lineaments of biopower: The bureaucratic and technological paradoxes of Aadhaar. *South Asia: Journal of South Asian Studies*, 42(3), 606–611.

Browne, S. (2015). *Dark Matters: On the Surveillance of Blackness*. New York: Duke University Press.

Burt, C. (2023). History repeats with Kenyan High Court blocking Maisha Namba for lack of DPIA. Biometric Update. https://www.biometricupdate.com/202312/history-repeats-with-kenyan-high-court-blocking-maisha-namba-for-lack-of-dpia

Business Quant (n.d.). *Palantir's Revenue by Segment 2019–2023*. https://businessqu ant.com/palantir-revenue-by-segment

Carswell, G., & De Neve, G. (2020). Paperwork, patronage, and citizenship: The materiality of everyday interactions with bureaucracy in Tamil Nadu, India. *The Journal of the Royal Anthropological Institute*, 26(3), 495–514.

Carswell, G., & De Neve, G. (2022). Transparency, exclusion and mediation: How digital and biometric technologies are transforming social protection in Tamil Nadu, India. *Oxford Development Studies*, 50(2), 126–141.

Centre for Human Rights and Global Justice (2021). *Chased Away and Left to Die: How a National Security Approach to Uganda's National Digital ID Has Led to Wholesale Exclusion of Women and Older Persons*. New York University. https://chrgj.org/wp-content/uploads/2021/06/CHRGJ-Report-Chased-Away-and-Left-to-Die.pdf

Centre for Human Rights and Global Justice (2022). *Paving a Digital Road to Hell: A Primer on the Role of the World Bank and Global Networks in Promoting Digital ID*. New York University. https://chrgj.org/wp-content/uploads/2022/06/Report_ Paving-a-Digital-Road-to-Hell.pdf

Centre for Human Rights and Global Justice (2023). *Prominent Human Rights Expert Admitted as Amicus Curiae in Groundbreaking Legal Challenge to Ugandan National Digital ID System*. https://chrgj.org/2023/03/24/prominent-human-rights-expert-admitted-as-amicus-curiae-in-groundbreaking-legal-challenge-to-ugandan-natio nal-digital-id-system/

Chaudhuri, B. (2021). Distant, opaque and seamful: Seeing the state through the workings of Aadhaar in India. *Information Technology for Development*, 27(1), 37–49.

Chaudhuri, B. (2022). Programmed welfare: An ethnographic account of algorithmic practices in the Public Distribution System in India. *New Media & Society*, 24(4), 887–902.

Cheesman, M. (2022a). *Infrastructure Justice and Humanitarianism: Blockchain's Promises in Practice*. Doctoral dissertation, University of Oxford.

Cheesman, M. (2022b). Self-sovereignty for refugees? The contested horizons of digital identity. *Geopolitics*, 27(1), 134–159.

Cheesman, M. (2022c). Blockchain for refugees. *Points*, https://points.datasociety.net/ blockchain-for-refugees-a46b41594eee

Cioffi, K. (2023). *Mass Exclusion from the National ID System in Uganda*. Open Global Rights. https://www.openglobalrights.org/human-rights-gateway-gatekeeper-dig ital-ids-uganda/

CIPESA (2023). *Uganda's Digital ID System Hinders Citizens' Access to Social Services*. https://cipesa.org/2023/10/ugandas-digital-id-system-hinders-citizens-access-to-social-services/#:~:text=Uganda's%20digital%20ID%20system%20is,print%2C% 20and%20eye%20scan%20information

Coppi, G., Jimenez, R. M., & Kyriazi, S. (2021). Explicability of humanitarian AI: A matter of principles. *Journal of International Humanitarian Action*, 6, 1–22.

Corbridge, S., Williams, G., Srivastava, M., & Véron, R. (2005). *Seeing the State: Governance and Governmentality in India*. Cambridge: Cambridge University Press.

Costanza-Chock, S. (2020). *Design Justice: Community-Led Practices to Build the Worlds We Need*. New York: The MIT Press.

Couldry, N., & Mejias, U. A. (2019). Data colonialism: Rethinking big data's relation to the contemporary subject. *Television & New Media*, 20(4), 336–349.

Crenshaw, K. (1991). Race, gender, and sexual harassment. *Southern California Law Review*, 65, 1467–11476.

Cristiano, F. (2022). *The Blurring Politics of Cyber Conflict: A Critical Study of the Digital in Palestine and beyond*. Lund: MediaTryck.

Cristiano, F. (2019). Internet access as human right: A dystopian critique from the Occupied Palestinian Territory. In Blouin-Genest, G., Doran, M.-C., & Paquerot, S. (Eds.), *Human Rights as Battlefields: Changing Practices and Contestations* (pp. 249–268). London: Springer International Publishing.

Cuesta, J., & Pico, J. (2020). The gendered poverty effects of the COVID-19 pandemic in Colombia. *European Journal of Development Research*, 32, 1558–1591.

Cunha, P. R. D., Soja, P., & Themistocleous, M. (2021). Blockchain for development: A guiding framework. *Information Technology for Development*, 27(3), 417–438.

Cusumano, M. A., Gawer, A., & Yoffie, D. B. (2019). *The Business of Platforms: Strategy in the Age of Digital Competition, Innovation, and Power*. New York: Harper Business.

Dahan, M., & Gelb, A. (2015). *The Role of Identification in the Post-2015 Development Agenda*. World Bank, Open Knowledge Repository. https://openknowledge.world bank.org/bitstream/handle/10986/22513/The0role0of0id050development0agenda. pdf

Dencik, L., Hintz, A., Redden, J., & Treré, E. (2022). *Data Justice*. London: SAGE.

Devereux, S. (2016). Is targeting ethical? *Global Social Policy*, 16(2), 166-181.

Devereux, S., & Sabates-Wheeler, R. (2004). *Transformative Social Protection*. Working Paper 232, Brighton: Institute of Development Studies, University of Sussex.

Devereux, S., & Vincent, K. (2010). Using technology to deliver social protection: Exploring opportunities and risks. *Development in Practice*, 20(3), 367–379.

Drèze, J., & Khera, R. (2015). Understanding leakages in the public distribution system. *Economic and Political Weekly*, 50(7), 39–42.

Drèze, J., & Khera, R. (2017). Recent social security initiatives in India. *World Development*, 98, 555–572.

Drèze, J., Khalid, N., Khera, R., & Somanchi, A. (2017). Pain without gain? Aadhaar and food security in Jharkhand. *Economic and Political Weekly*, 52(50), 50–60.

Drèze, J., Khera, R., & Somanchi, A. (2020a). *Balancing Corruption and Exclusion: A Rejoinder*. Ideas for India. https://www.ideasforindia.in/topics/poverty-inequality/ balancing-corruption-and-exclusion-a-rejoinder.html

Drèze, J., Khera, R., & Somanchi, A. (2020b). *Balancing Corruption and Exclusion: Comment on the Response*. Ideas for India. https://www.ideasforindia.in/topics/ poverty-inequality/balancing-corruption-and-exclusion-comment-on-the-response.html

Dutta, B., & Ramaswami, B. (2001). Targeting and efficiency in the Public Distribution System: Case of Andhra Pradesh and Maharashtra. *Economic and Political Weekly*, 36(18), 1524–1532.

El Tiempo (2020). *Inconsistencias en base de datos de Ingreso Solidario: Registraduria*. https://www.eltiempo.com/politica/las-inconsistencias-en-las-bases-de-ingreso-solidario-484754

Elyachar, J. (2010). Phatic labor, infrastructure, and the question of empowerment in Cairo. *American Ethnologist*, 37(3), 452–464.

Eubanks, V. (2018). *Automating Inequality: How High-Tech Tools Profile, Police, and Punish the Poor*. New York: St. Martin's Press.

EuroMed Rights (2023). *Artificial Intelligence: The New Frontier of the EU Border's Externalisation Strategy*. EuroMed Rights, https://euromedrights.org/wp-content/uploads/2023/07/Euromed_AI-Migration-Report_EN-1.pdf

Fassin, D. (2005). Compassion and repression: The moral economy of immigration policies in France. *Cultural Anthropology*, 20(3), 362–387.

Feuer, W. (2020). Palantir CEO Alex Karp defends his company's relationship with government agencies. *CNBC*. https://www.cnbc.com/2020/01/23/palantir-ceo-alex-karp-defends-his-companys-work-for-the-government.html

Floridi, L. (2023). *The Ethics of Artificial Intelligence: Principles, Challenges, and Opportunities*. London: Oxford University Press.

Fraser, N. (2008). Abnormal justice. *Critical Inquiry*, 34(3), 393–422.

Fussy, P. (2021). Liberating COVID-19 data with volunteers in Brazil. In Milan, S., Treré, E., & Masiero, S. (Eds.), *COVID-19 from the Margins. Pandemic Invisibilities, Policies and Resistance in the Datafied Society*. Amsterdam: Institute of Network Cultures, pp. 241–245.

Gelb, A., & Clark, J. (2013). *Identification for Development: The Biometrics Revolution*. Center for Global Development Working Paper, p. 315.

Gill (2020). *Explained: In Kenya's Digital ID System, Echoes of India's Aadhaar*. https://indianexpress.com/article/explained/in-kenyas-digital-id-system-echoes-of-indias-aadhaar-6244643

Global Freedom of Expression (n.d.) *Puttaswamy V. Union of India (II)*. https://globalfreedomofexpression.columbia.edu/cases/puttaswamy-v-union-of-india-ii/

Gonzalez, B. (2023). Nubian Rights Forum urges equitable rollout to Kenya's UPI system. Biometric Update. https://www.biometricupdate.com/202306/nubian-rights-forum-urges-equitable-rollout-to-kenyas-upi-system

Gopakumar, K. (2007). E-governance services through telecentres-role of human intermediary and issues of Trust. In *2006 International Conference on Information and Communication Technologies and Development* (pp. 131–142). Berkeley (CA), USA, IEEE.

Government of India (1968). *Report of the Jha Committee on Foodgrain Prices for 1964–65*. New Delhi: India Foodgrain Prices Committee, Department of Agriculture.

Government of India (2015). *Wiping every tear from every eye: The JAM Trinity number solution. Economic Survey 2015–2016*. New Delhi: Government of India. http://indiabudget.nic.in/es2014-15/echapvol1-03.pdf

Gulati, A., & Saini, S. (2015). *Leakages from Public Distribution System (PDS) and the Way Forward*. Working Paper, No. 294, New Delhi: Indian Council for Research on International Economic Relations (ICRIER).

Haki na Sheria (2021). *Biometric Purgatory: How the Double Registration of Vulnerable Kenyan Citizens in the UNHCR Database has Left Them at Risk of Statelessness*. http://citizenshiprightsafrica.org/wp-content/uploads/2021/11/Haki-na-Sheria_Double-Registration_Nov2021.pdf

Haki na Sheria (2022). *Press Statement*, 18 January 2022. http://citizenshiprightsafrica.org/wp-content/uploads/2022/01/Haki-na-Sheria-Press-Release-18Jan2022.pdf

Haki na Sheria (2023). *Memorandum on Implementation of Digital ID. Civil Society Organisations Coalition Document*, 25 September 2023.

Haki na Sheria (n.d.). *Our Story*. https://hakinasheria.org

Hayes De Kalaf, E. (2019). Making foreign: Legal identity, social policy and the contours of belonging in the contemporary Dominican Republic. In Cruz-Martínez, G. (Ed.), *Welfare and Social Protection in Contemporary Latin America* (pp. 101–117). London: Routledge.

Hayes de Kalaf, E. (2021). *Legal Identity, Race and Belonging in the Dominican Republic: From Citizen to Foreigner*. London: Anthem Press.

Heeks, R. (2002). Information systems and developing countries: Failure, success, and local improvisations. *The Information Society*, 18(2), 101–112.

Heeks, R. (2022). Digital inequality beyond the digital divide: Conceptualizing adverse digital incorporation in the global South. *Information Technology for Development*, 28(4), 688–704.

Heeks, R., & Renken, J. (2018). Data justice for development: What would it mean? *Information Development*, 34(1), 90–102.

Heeks, R., & Shekhar, S. (2019). Datafication, development and marginalised urban communities: An applied data justice framework. *Information, Communication & Society*, 22(7), 992–1011.

Hersey, F. (2022). ID is no longer just about identity: Day 1 of ID4Africa 2022. Biometric Update. https://www.biometricupdate.com/202206/id-is-no-longer-just-about-identity-day-1-of-id4africa-2022

Hersey, F. (2023). Oxford Refugee Studies Centre: Mapping the biometrics of movement into and around Europe. Biometric Update. https://www.biometricupdate.com/202301/oxford-refugee-studies-centre-mapping-the-biometrics-of-movement-into-and-around-europe

Hindustan Times (2017). Cash transfer gets thumbs down, less than 60% cash reaches beneficiaries. https://www.hindustantimes.com/india-news/cash-transfer-gets-thumbs-down-less-than-60-cash-reaches-beneficiaries/story-Aw9K2movh4xprVAcE71aQP.html

Hindustan Times (2022). Reforming the PDS for better nutrition. https://www. hindustantimes.com/ht-insight/public-health/reforming-the-pds-for-better-nutrition-101667826489737.html

Human Rights Watch (2023). *World Bank/Jordan: Poverty Targeting Algorithms Harm Rights*. https://www.hrw.org/news/2023/06/13/world-bank/jordan-poverty-targeting-algorithms-harm-rights

Hundal, H. S., Janani, A. P., & Chaudhuri, B. (2020). A conundrum of efficiency and inclusion: Aadhaar and fair-price shops. *Economic and Political Weekly*. https://www.epw.in/engage/article/conundrum-efficiency-and-inclusion-aadhaar-and

Hvistendahl, M. (2021). How the LAPD and Palantir use data to justify racist policing. *The Intercept*. https://theintercept.com/2021/01/30/lapd-palantir-data-driven-policing/

Iazzolino, G. (2021). Infrastructure of compassionate repression: Making sense of biometrics in Kakuma refugee camp. *Information Technology for Development*, 27(1), 111–128.

Iazzolino, G., & Stremlau, N. (2024). AI for social good and the corporate capture of global development. *Information Technology for Development*, 1–18.

ID4D (2019). *ID4D's Practitioner Guide*. https://documents.worldbank.org/en/publication/documents-reports/documentdetail/248371559325561562/id4d-practitioner-s-guide

Iliadis, A., & Acker, A. (2022). The seer and the seen: Surveying Palantir's surveillance platform. *The Information Society*, 38(5), 334–363.

ILO (2016). *Janani Suraksha Yojana Guidelines for Implementation*. https://webapps. ilo.org/dyn/travail/docs/683/JananiSurakshaYojanaGuidelines/MinistryofHealth andFamilyWelfare.pdf

ILO (2017). *SISBEN: A Unified Vulnerability Assessment and Identification System for Social Assistance*. https://www.ilo.org/wcmsp5/groups/public/—dgreports/—integration/documents/publication/wcms_568689.pdf

Isaac, T. M., & Tharakan, P. K. (1995). Kerala: Towards a new agenda. *Economic and Political Weekly*, 30(31/32), 1993–2004.

Justice Wadhwa Committee on Public Distribution System. (2010). *Report on the State of Kerala*. New Delhi: Central Vigilance Committee on the Public Distribution System.

Kenya News Agency (2023). Plans for Kenyans to get new identifier by June 1. https://www.pd.co.ke/news/plans-for-kenyans-to-get-new-identifier-by-june-1-180775/

Khera, R. (2011a). India's Public Distribution System: Utilisation and impact. *Journal of Development Studies*, 47(7), 1038–1060.

Khera, R. (2011b). Trends in diversion of grain from the Public Distribution System. *Economic and Political Weekly*, 46(21), 106–114.

Khera, R. (2011c). The UID project and welfare schemes. *Economic and Political Weekly*, 46(9), 38–43.

Khera, R. (2014). Cash vs. in-kind transfers: Indian data meets theory. *Food Policy*, 46, 116–128.

Khera, R. (2017). Impact of Aadhaar on welfare programmes. *Economic and Political Weekly*, 52(50), 61–70.

Khera, R. (2018). Smarter than Aadhaar: Govt's Insistence on Disruptive Option Is Bewildering. *Business Standard*, 14 March 2018. https://www.business-stand ard.com/article/opinion/how-successfully-last-mile-authentication-has-recorded-pds-118031301260_1.html

Khera, R. (2019). *Dissent on Aadhaar: Big Data Meets Big Brother.* New Delhi: Orient Blackswan.

Kidd, S. (2017). *Uganda's Senior Citizens' Grant: A Success Story from the Heart of Africa.* https://www.developmentpathways.co.uk/wp-content/uploads/2018/06/Summary -of-Uganda-Senior-Citizens-Grant-evaluation.pdf

Kitchin, R. (2014). Big Data, new epistemologies and paradigm shifts. *Big Data & Society*, 1(1), 1–18.

Kitchin, R. (2016) The ethics of smart cities and urban science. *Philosophical Transactions of the Royal Society A*, 374(2083), 1–15.

Kopytoff, I. (1986). The cultural biography of things: Commoditization as process. In Appadurai, A. (Ed.), *The Social Life of Things. Commodities in Cultural Perspective* (pp. 64–91). Cambridge: Cambridge University Press.

Krishna, S. (2021). Digital identity, datafication and social justice: Understanding Aadhaar use among informal workers in south India. *Information Technology for Development*, 27(1), 67–90.

Krishnakumar, R. (2000). Public distribution system: A system in Peril. *Frontline*, 17(19), 16–29.

Krishnan, V. (2023). Structures behind numbers: Critically examining the 'credibility revolution' and 'evidence-based policy'. *Economic and Political Weekly*, https:// www.epw.in/engage/article/structures-behind-numbers-critically-examining

Kuriyan, R., & Ray, I. (2009). Outsourcing the state? Public–private partnerships and information technologies in India. *World Development*, 37(10), 1663–1673.

Larkin, B. (2013). The politics and poetics of infrastructure. *Annual Review of Anthropology*, 42(1), 327–343.

Latonero, M., Hiatt, K., Napolitano, A., Clericetti, G., & Penagos, M. (2019). Digital identity in the migration & refugee context: Italy case study. https://datasociety. net/library/digital-identity-in-the-migration-refugee-context/

López, J. (2021a). The case of the Solidarity Income in Colombia: The experimentation with data on social policy during the pandemic. In Milan, S., Treré, E., & Masiero, S. (Eds.), *COVID-19 from the Margins: Pandemic Invisibilities, Policies and Resistance in the Datafied Society* (pp. 126–128). Amsterdam: Institute of Network Cultures.

López, J. (2021b). *Experimentando con la pobreza: El Sisbén y los proyectos de analítica de datos en Colombia.* https://web.karisma.org.co/wp-content/uploads/download-manager-files/Experimentando%20con%20la%20pobreza.pdf

López, J. (2022). *Data for Dignity: Requirements for the Implementation of Data Systems for Social Programs in Colombia.* Tilburg University, https://research.tilburguniversity. edu/en/publications/data-for-dignity-requirements-for-the-implementation-of-data-syst

Lowe, C. (2022). *The Digitalisation of Social Protection before and since the Onset of COVID-19: Opportunities, Challenges and Lessons*, Working paper. ODI. https://cdn.odi.org/media/documents/ODI_Working_paper_Digitalisation_of_social_protection.pdf

Lyon, D. (2001). *Surveillance Society: Monitoring Everyday Life*. London: McGraw-Hill Education.

Lyon, D. (2007). *Surveillance Studies: An Overview*. London: Polity.

Lyon, D. (2009). *Identifying Citizens: ID Cards as Surveillance*. London: Polity.

Madianou, M. (2019). The biometric assemblage: Surveillance, experimentation, profit, and the measuring of refugee bodies. *Television & New Media*, 20(6), 581–599.

Madon, S. (2005). Governance lessons from the experience of telecentres in Kerala. *European Journal of Information Systems*, 14, 401–416.

Madon, S., & Schoemaker, E. (2021). Digital identity as a platform for improving refugee management. *Information Systems Journal*, 31(6), 929–953.

Madon, S., Ranjini, C. R., & Anantha Krishnan, R. K. (2022). Aadhaar and social assistance programming: Local bureaucracies as critical intermediary. *Information Technology for Development*, 28(4), 705–720.

Manby, B. (2021). The Sustainable Development Goals and 'legal identity for all': 'First, do no harm'. *World Development*, 139, 105343.

Manorama (2017). Colour-coding of ration cards. https://www.onmanorama.com/news/kerala/2017/05/30/colour-coded-ration-cards-kerala.html

Marabelli, M. (2024). *AI, Ethics, and Discrimination in Business: The DEI Implications of Algorithmic Decision-Making*. New York: Palgrave MacMillan.

Martin, A. (2019). Mobile money platform surveillance. *Surveillance and Society*, 17(1/2), 213–222.

Martin, A. (2021). Aadhaar in a box? Legitimizing digital identity in times of crisis. *Surveillance and Society*, 19(1), 104–108.

Martin, A. (2023). Aidwashing surveillance: Critiquing the corporate exploitation of humanitarian crises. *Surveillance and Society*, 21(1), 96–102.

Martin, A., & Taylor, L. (2021a). Exclusion and inclusion in identification: Regulation, displacement and data justice. *Information Technology for Development*, 27(1), 50–66.

Martin, A., & Taylor, L. (2021b). Give us your poor, your unidentified masses. *Global Data Justice*, https://globaldatajustice.org/2021-09-29-identity-week-2021/

Martin, A., Sharma, G., Peter de Souza, S., Taylor, L., van Eerd, B., McDonald, S. M., & Dijstelbloem, H. (2023). Digitisation and sovereignty in humanitarian space: Technologies, territories and tensions. *Geopolitics*, 28(3), 1362–1397.

Martins, B. O. (2023). Security knowledges: Circulation, control, and responsible research and innovation in EU border management. *Science as Culture*, 32(3), 435–459.

Martins, B. O., & Jumbert, M. G. (2022). EU Border technologies and the co-production of security 'problems' and 'solutions'. *Journal of Ethnic and Migration Studies*, 48(6), 1430–1447.

Martins, B. O., Lidén, K., & Jumbert, M. G. (2022). Border security and the digitalisation of sovereignty: Insights from EU borderwork. *European Security*, 31(3), 475–494.

Masiero, S. (2011). Financial vs social sustainability of telecentres: Mutual exclusion or mutual reinforcement? *The Electronic Journal on Information Systems in Developing Countries*, 45(1), 1–23.

Masiero, S. (2012). *Transforming State-Citizen Relations in Food Security Schemes: The Computerized Ration Card Management System in Kerala*. Working paper, CDS Trivandrum. http://14.139.171.199:8080/xmlui/handle/123456789/56

Masiero, S. (2014). *Imagining the State through Digital Technologies: A Case of State-Level Computerization in the Indian Public Distribution System*. PhD dissertation, London School of Economics and Political Science (LSE), London.

Masiero, S. (2015a). Redesigning the Indian food security system through e-governance: The case of Kerala. *World Development*, 67, 126–137.

Masiero, S. (2015b). Will the JAM trinity dismantle the PDS? *Economic and Political Weekly*, 50(45), 21–23.

Masiero, S. (2020). Biometric infrastructures and the Indian public distribution system. *South Asia Multidisciplinary Academic Journal*, 23(3), 1–18.

Masiero, S. (2023). Dark side of IT: A misleading expression? *The Electronic Journal on Information Systems in Developing Countries*, 1–14.

Masiero, S., & Arvidsson, V. (2021). Degenerative outcomes of digital identity platforms for development. *Information Systems Journal*, 31(6), 903–928.

Masiero, S., & Bailur, S. (2021). Digital identity for development: The quest for justice and a research agenda. *Information Technology for Development*, 27(1), 1–12.

Masiero, S., & Das, S. (2019). Datafying anti-poverty programmes: Implications for data justice. *Information, Communication & Society*, 22(7), 916–933.

Masiero, S., & Prakash, A. (2015) The politics of anti-poverty artefacts: Lessons from computerisation of the food security system in Karnataka. In *7th International Conference on Information and Communication Technologies for Development (ICTD 2015)*. Singapore, 15–18 May 2015.

Masiero, S., & Prakash, A. (2019) ICT in social protection schemes: Deinstitutionalising subsidy-based welfare programmes. *Information Technology & People*, 33(4), 1255–1280.

Masiero, S., & Shakthi, S. (2020). Grappling with Aadhaar: Biometrics, social identity and the Indian state. *South Asia Multidisciplinary Academic Journal*, 23, 1–8.

Mathrubhumi (2017). New Kerala ration cards in 4 colours, to be distributed from June. 29 May. https://englisharchives.mathrubhumi.com/news/kerala/new-kerala-ration-cards-in-4-colours-to-be-distributed-from-june-1.1973830

Mayer-Schönberger, V., & Cukier, K. (2013). *Big Data: A Revolution that Will Transform How We Live, Work, and Think*. London: Houghton Mifflin Harcourt.

McCall-Smith, K. (2023). Good better best? Human rights impact assessment in crisis lawmaking. *The International Journal of Human Rights*, 27(9–10), 1326–1344.

McDonald, A. Uganda's digital ID achievements, challenges and prospects. Biometric Update. https://www.biometricupdate.com/202201/ugandas-digital-id-achieveme nts-challenges-and-prospects

McDonald, A. (2023). MOSIP-based legal, digital ID enrolments hit 100M milestone. Biometric Update. https://www.biometricupdate.com/202312/mosip-based-legal-digital-id-enrollments-hit-100m-milestone

Melguizo, T., Sanchez, F., & Velasco, T. (2016). Credit for low-income students and access to and academic performance in higher education in Colombia: A regression discontinuity approach. *World Development*, 80, 61–77.

Milan, S., & Van der Velden, L. (2016). The alternative epistemologies of data activism. *Digital Culture & Society*, 2(2), 57–74.

Milan, S., Treré, E., & Masiero, S. (2021). *COVID-19 from the Margins: Pandemic Invisibilities, Policies and Resistance in the Datafied Society*. Amsterdam: Institute of Network Cultures.

Mir, U. B., Kar, A. K., Dwivedi, Y. K., Gupta, M. P., & Sharma, R. S. (2020). Realizing digital identity in government: Prioritizing design and implementation objectives for Aadhaar in India. *Government Information Quarterly*, 37(2), 101442.

Mooij, J. E. (1994). Public distribution system as safety net: Who is saved? *Economic and Political Weekly*, 29(3), 119–126.

Mooij, J. (1998). Food policy and politics: The political economy of the public distribution system in India. *Journal of Peasant Studies*, 25(2), 77–101.

Mooij, J. (1999). Food policy in India: The importance of electoral politics in policy implementation. *Journal of International Development*, 11(4), 625–636.

Mukhopadhyay, S., Bouwman, H., & Jaiswal, M. P. (2019). An open platform centric approach for scalable government service delivery to the poor: The Aadhaar case. *Government Information Quarterly*, 36(3), 437–448.

Murakami Wood, D., & Firmino, R. (2009). Empowerment or repression? Opening up questions of identification and surveillance in Brazil through a case of 'identity fraud'. *Identity in the Information Society*, 2, 297–317.

Muralidharan, K., Niehaus, P., & Sukhtankar, S. (2020). *Balancing Corruption and Exclusion: Incorporating Aadhaar into PDS*. Ideas for India, https://www.ideasforindia. in/topics/povertyinequality/balancing-corruption-and-exclusion-incorporating-aadhaar-into-pds.html

Mutung'u, G., & Rutenberg, I. (2020). Digital ID and risk of statelessness. *Statelessness & Citizenship Review*, 2: 348–358.

Mutung'u, G. (2021). *Digital Identity in Kenya*. https://researchictafrica.net/wp/wp-content/uploads/2021/11/Kenya_1.11.21.pdf

Nair, E. C. (2000). Interview: 'Centre should change its policy'. *Frontline*, 17(19), 16–29.

Namati (2021). *Impact Report*. https://namati.org/our-impact/impact-report-2021/ citizenship/

Nation (2023). High Court puts the brakes on *Kindiki*'s plan to introduce Maisha Namba. https://nation.africa/kenya/news/high-court-puts-the-brakes-on-kindiki-s-plan-to-introduce-maisha-namba-4454474

Neves, J. A., Vasconcelos, F. D. A. G. D., Machado, M. L., Recine, E., Garcia, G. S., & Medeiros, M. A. T. D. (2022). The Brazilian cash transfer program (Bolsa Família): A tool for reducing inequalities and achieving social rights in Brazil. *Global Public Health*, 17(1), 26–42.

Newell, B. C. (Ed.). (2020). *Police on Camera: Surveillance, Privacy, and Accountability*. London: Routledge.

NFSA (2013). *National Food Security Act*. https://nfsa.gov.in/portal/nfsa-act

Nilekani, N. (2013). *The Science of Delivering On-line IDs for a Billion People: The Aadhaar Experience*. https://www.worldbank.org/en/news/video/2013/04/24/the-science-of-delivering-on-line-ids-for-a-billion-people-the-aadhaar-experience

Nilekani, N., & Shah, V. (2016). *Rebooting India: Realizing a Billion Aspirations*. London: Penguin.

Nyst, C., Makin, P., Pannifer, S., & Whitley, E. (2016). *Digital Identity: Issue Analysis: Executive Summary*. Consult Hyperion.

Office of the Auditor General (2022). *A Value for Money Audit Report on Management of Senior Citizens Grant by the Expanding Social Protection Programme under the Ministry of Gender*, Labour and Social Development. https://www.oag.go.ug/reports/1264

Pal, J., Nedevschi, S., Patra, R. K., & Brewer, E. A. (2006). A multidisciplinary approach to open access village telecenter initiatives: The case of akshaya. *E-Learning and Digital Media*, 3(3), 291–316.

Palantir (n.d.). *Gotham Platform*. https://www.palantir.com/platforms/gotham/

Palantir (n.d.). *Palantir Foundry*. https://www.palantir.com/platforms/foundry/

PalTel Group (2023). *Paltel Group's Response to Access Now and Human Rights Watch's Letter*. https://www.accessnow.org/press-release/paltel-group-response-letter-access-now-hrw/

Parker, B. (2019). New UN deal with data mining firm Palantir raises protection concerns. https://www.thenewhumanitarian.org/news/2019/02/05/un-palantir-deal-data-mining-protection-concerns-wfp

Parmiggiani, E., Østerlie, T., & Almklov, P. G. (2022). In the backrooms of data science. *Journal of the Association for Information Systems*, 23(1), 139–164.

Pelizza, A. (2020). Processing alterity, enacting Europe: Migrant registration and identification as co-construction of individuals and polities. *Science, Technology & Human Values*, 45(2), 262–288.

Pelizza, A., & Loschi, C. (2023). Telling 'more complex stories' of European integration: How a sociotechnical perspective can help explain administrative continuity in the Common European Asylum System. *Journal of European Public Policy*, 1–22.

Pelizza, A, Milan, S., & Lausberg, Y. (2021). The dilemma of undocumented migrants invisible to COVID-19. In Milan, S., Treré, E., & Masiero, S. (Eds.), *COVID-19 from the Margins: Pandemic Invisibilities, Policies and Resistance in the Datafied Society.* (pp. 70–78). Amsterdam: Institute of Network Cultures.

Polito, C. & Alaimo, C. (2023). The politics of biometric technologies: Borders control and the making of data citizens in Africa. In *European Conference of Information Systems (ECIS)*. Kristiansand, 13–16 June 2023.

Prabhakar, T. (2017). New ration policy, same old problems. *New Indian Express*, 15 October. https://www.newindianexpress.com/states/tamil-nadu/2017/oct/15/new-ration-policy-same-old-problems-1674554.html

Prakash, A., & Masiero, S. (2015). Does computerisation reduce PDS leakage? Lessons from Karnataka. *Economic and Political Weekly*, 50(50), 77–81.

Prieto, A. M. (2021). *Ingreso Solidario: A New Generation of G2P Payments in Colombia.* https://thedocs.worldbank.org/en/doc/e33a6191df6cb47288574167b86efa50-0380022021/original/CKEx-May-10-Innovations-in-SP-Delivery-Colombia.pdf

Privacy International (2018). *The Humanitarian Metadata Problem – Doing No Harm in the Digital Era.* https://privacyinternational.org/report/2509/humanitarian-meta data-problem-doing-no-harm-digital-era

Privacy International (2019a). *Understanding Identity Systems Part 1: Why ID?* https://privacyinternational.org/explainer/2669/understanding-identity-systems-part-1-why-id

Privacy International (2019b). *Understanding Identity Systems Part 2: Discrimination and Identity.* https://www.privacyinternational.org/explainer/2670/understanding-identity-systems-part-2-discrimination-and-identity

Privacy International (2021). *Digital National ID Systems: Ways, Shapes and Forms.* https://privacyinternational.org/long-read/4656/digital-national-id-systems-ways-shapes-and-forms

Privacy International (2022). *Data Protection Impact Assessments and ID Systems: The 2021 Kenyan Ruling on Huduma Namba.* https://privacyinternational.org/news-analysis/4778/data-protection-impact-assessments-and-id-systems-2021-kenyan-ruling-huduma

Privacy International (n.d.) *Impact.* https://privacyinternational.org/impact

Puri, R. (2012). Reforming the public distribution system: Lessons from Chhattisgarh. *Economic and Political Weekly*, 47(5), 21–23.

Radhakrishna, R., & Subbarao, K. (1997). *India's Public Distribution System: A National and International Perspective.* Washington, DC: World Bank Publications.

Raghavan, M. (2021). *Transaction Failure Rates in the Aadhaar-Enabled Payment System.* Dvara Research. https://www.dvara.com/research/wp-content/uploads/2020/05/Transaction-failure-rates-in-the-Aadhaar-enabled-Payment-System-Urgent-issues-for-consideration-and-proposed-solutions.pdf

Raghunandan, P.M. (2013). Ensuring fairness at shops with e-machines. *Deccan Herald.* https://www.deccanherald.com/content/303290/ensuring-fairness-shops-e-machines.html

Ramakumar, R. (2010). The unique ID project in India: A skeptical note. In *International Conference on Ethics and Policy of Biometrics* (pp. 154–168). Berlin, Heidelberg, Springer.

Ramaswami, B., & Balakrishnan, P. (2002). Food prices and the efficiency of public intervention: The case of the public distribution system in India. *Food Policy*, 27(5), 419–436.

Rankin, J. (2021). EU 'seeking to turn migrant database into mass surveillance tool'. https://www.theguardian.com/world/2021/sep/08/eu-seeking-to-turn-migrant-database-into-mass-surveillance-tool

Rao, U. (2013). Biometric marginality: UID and the shaping of homeless identities in the city. *Economic and Political Weekly*, 48(13), 71–77.

Rao, U. (2017). Writing, typing and scanning: Distributive justice and the politics of visibility in the era of e-governance. In *Media as Politics in South Asia* (pp. 127–140). London: Routledge.

Rao, U., & Nair, V. (2019). Aadhaar: Governing with biometrics. *South Asia: Journal of South Asian Studies*, 42(3), 469–481.

Rawls, J. (1971). *A Theory of Justice*. Oxford: Oxford University Press.

Renzi, A, & Langlois, G. (2015). Data activism. In Elmer, G., Langlois, G., & Redden, J. (Eds.), *Compromised Data: From Social Media to Big Data* (pp. 202–225). London: Bloomsbury.

Responsible Data (2019). *Open Letter to WFP re: Palantir Agreement.* https://responsibledata.io/2019/02/08/open-letter-to-wfp-re-palantir-agreement

Reuters (2022). Uganda sued over digital ID system that excludes millions. https://news.trust.org/item/20220513155511-l53t0/

Right to Food India (2016a). *National Food Security Act: A Primer.* https://drive.google.com/file/d/1NAAyjKQZ0axSddWdoVMhzIlhKunwOs9a/view

Right to Food India (2016b). *Anganwadis for All: A Primer.* https://drive.google.com/file/d/1TqQZ_jQ9k5If0gyckWZmzaf5AzCoHp-Q/view

Right to Food India (n.d.). *About.* https://www.righttofoodcampaign.in/about

Roberts, T., Mohamed Ali, A., Karekwaivanane, G., Msonza, N., Phiri, S., Nanfuka, J., & Ndongmo, K. (2021). *Digital Rights in Closing Civic Space: Lessons from Ten African Countries.* Institute of Development Studies. https://opendocs.ids.ac.uk/opendocs/handle/20.500.12413/15964

Roman, R., & Colle, R. D. (2001). *Sustaining the Community Telecenter Movement.* New York: Cornell University.

Salo, M., Pirkkalainen, H., Chua, C. E. H., & Koskelainen, T. (2022). Formation and mitigation of technostress in the personal use of IT. *MIS Quarterly*, 46, 1–36.

Sandvik, K.B. (2023). *Humanitarian Extractivism: The Digital Transformation of Aid.* London: Manchester University Press.

Schoemaker, E., Baslan, D., Pon, B., & Dell, N. (2021). Identity at the margins: Data justice and refugee experiences with digital identity systems in Lebanon, Jordan, and Uganda. *Information Technology for Development*, 27(1), 13–36.

Schoemaker, E., Martin, A., & Weitzberg, K. (2023). Digital identity and inclusion: Tracing technological transitions. *Georgetown Journal of International Affairs*, 24(1), 36–45.

Schrock, A. R. (2016). Civic hacking as data activism and advocacy: A history from publicity to open government data. *New Media & Society*, 18(4), 581–599.

Scott, J. C. (1998). *Seeing Like a State: How Certain Schemes to Improve the Human Condition Have Failed*. New Haven: Yale University Press.

Sen, A., & Himanshu (2011). Why not a universal food security legislation? *Economic and Political Weekly*, 46(12), 38–47.

Shakthi, S. (2020). Crafting 'integrity': The implications of authentication through unique identification databases. *South Asia Multidisciplinary Academic Journal*, 23, 1–18.

Sharon, T. (2021). Blind-sided by privacy? Digital contact tracing, the Apple/Google API and big tech's newfound role as global health policy makers. *Ethics and Information Technology*, 23(1), 45–57.

Sherman, N. (2020). Palantir: The controversial data firm now worth £17bn. *BBC News*. https://www.bbc.com/news/business-54348456

Singh, S. (2019). Death by digital exclusion? On faulty public distribution system in Jharkhand. *The Hindu*. https://www.thehindu.com/news/national/other-states/death-by-digital-exclusion/article28414768.ece

Singh, R. P. (2020). *Seeing like an Infrastructure: Mapping Uneven State-Citizen Relations in Aadhaar-Enabled Digital India*. PhD thesis, Ithaca: Cornell University.

Sinha, D. (2015). Cash for food—A misplaced idea. *Economic and Political Weekly*, 50(16), 17–20.

Solanki, A. (2019). Management of performance and performance of management: Getting to work on time in the Indian bureaucracy. *South Asia: Journal of South Asian Studies*, 42(3), 588–605.

Srinivasan, J. (2022). *The Political Lives of Information: Information and the Production of Development in India*. New York: MIT Press.

Srinivasan, J., & Johri, A. (2013). Creating machine readable men: legitimizing the 'Aadhaar' mega e-infrastructure project in India. In *Proceedings of the Sixth International Conference on Information and Communication Technologies and Development: Full Papers-Volume 1* (pp. 101–112). Cape Town, South Africa.

Sriraman, T. (2011). Revisiting welfare: Ration card narratives in India. *Economic and Political Weekly*, 46(38), 52–59.

Sriraman, T. (2013). Enumeration as pedagogic process: Gendered encounters with identity documents in Delhi's urban poor spaces. *South Asia Multidisciplinary Academic Journal*, 8, 1–18.

Sriraman, T. (2018). *In Pursuit of Proof: A History of Identification Documents in India*. New Delhi: Oxford University Press.

Star, S. L. (1999). The ethnography of infrastructure. *American Behavioral Scientist*, 43(3), 377–391.

Statewatch (2023). *Europe's Techno-Borders*. https://www.statewatch.org/publications/reports-and-books/europe-s-techno-borders/

Suchitra, M. (2004). *Undermining a Fine Public Distribution System in Kerala*. India Together. http://www.indiatogether.org/2004/jan/pov-keralapds.htm

Swaminathan, M. (2002). Excluding the needy: The public provisioning of food in India. *Social Scientist*, 30(3), 34–58.

Swaminathan, M. (2008). *Programmes to Protect the Hungry: Lessons from India*. Working Paper no.70 at United Nations Department for Economics and Social Affairs.

Tapasya, Sambhav, K., & Joshi, D. (2024). How an algorithm denied food to thousands of poor in India's Telangana. https://www.aljazeera.com/economy/2024/1/24/how-an-algorithm-denied-food-to-thousands-of-poor-in-indias-telangana

Tarafdar, M., Gupta, A., & Turel, O. (2013). The dark side of information technology use. *Information Systems Journal*, 23(3), 269–275.

Tarafdar, M., Gupta, A., & Turel, O. (2015). Special issue on 'dark side of information technology use': An introduction and a framework for research. *Information Systems Journal*, 25(3), 161–170.

Taylor, L. (2017). What is data justice? The case for connecting digital rights and freedoms globally. *Big Data & Society*, 4(2), 1–14.

Taylor, L., Martin, A., de Souza, S. P., & Lopez-Solano, J. (2023). Why are sector transgressions so hard to govern? Reflections from Europe's pandemic experience. *Information, Communication & Society*, 1–5.

Thales (n.d.). Eurodac: The European Union's first multinational biometric system. https://www.thalesgroup.com/en/markets/digital-identity-and-security/government/customer-cases/eurodac

Thatcher, J. (2014). Living on fumes: Digital footprints, data fumes, and the limitations of spatial big data. *International Journal of Communication*, 8, 19–38.

The Economic Times (2016). Government to plug PDS leakages with Aadhaar authentication. https://economictimes.indiatimes.com/news/politics-and-nation/government-to-plug-pds-leakages-with-aadhaar-authentication/articleshow/52473720.cmsl

The Engine Room (2023). *Biometrics in the Humanitarian Sector: A Current Look at Risks, Benefits and Organisational Policies*. https://www.theengineroom.org/biometrics-humanitarian-sector-2023/

The Hindu (2013). Border mafia behind illegal diversion of PDS rice. https://www.thehindu.com/news/cities/Thiruvananthapuram/border-mafia-behind-illegal-diversion-of-pds-rice/article4537084.ece

The Open Society Initiative (2021). *New Kenya High Court Judgment Sets Important Precedent for Digital ID Privacy Protections and Processes*. https://www.justiceinitiative.org/newsroom/new-kenya-high-court-judgment-sets-important-precedent-for-digital-id-privacy-protections-and-processes

The Wire (2016). Even in Delhi, Basing PDS on Aadhaar is Denying Many the Right to Food. *The Wire*. https://thewire.in/rights/right-to-food-how-aadhaar-in-pds-is-denying-rights

Treré, E. (2018). *Hybrid Media Activism: Ecologies, Imaginaries, Algorithms*. London: Routledge.

Tritah, A. (2003). *The Public Distribution System in India: Counting the Poor from Making the Poor Count. Groupe de Recherche en Economie Mathématique et Quantitative (GREMAQ)*. Université des Sciences Sociales, Toulouse.

UIDAI (2019). *UIDAI Annual Report 2018–2019*. https://uidai.gov.in/images/AADHAR _AR_2018_19_ENG_approved.pdf

UIDAI (2022). *State/UT Wise Aadhaar Saturation*. https://uidai.gov.in/images/State WiseAge_AadhaarSat_Rep_30112022_Projected-2022-Final.pdf

Umali-Deininger, D. L., & Deininger, K. W. (2001). Towards greater food security for India's poor: Balancing government intervention and private competition. *Agricultural Economics*, 25(2–3), 321–335.

UNDP (2020). *Coronavirus in Colombia: Vulnerability and Policy Options*. http://undp-rblac-cd19-pds-number11-en-colombia.pdf/36

UNHCR (2022). *Digital Payments to Refugees – A Pathway towards Financial Inclusion*. https://www.unhcr.org/media/39664

United Nations (2016). *Why the SDGs Matter*. https://www.un.org/sustainabledevelop ment/why-the-sdgs-matter/

Van Rossem, W., & Pelizza, A. (2022). The ontology explorer: A method to make visible data infrastructures for population management. *Big Data & Society*, 9(1), 1-14.

Veeraraghavan, R. (2021). *Patching Development: Information Politics and Social Change in India*. London: Oxford University Press.

Velez, V. O. (2020). *Not a Fairy Tale: Unicorns and Social Protection of Gig Workers in Colombia*. Working Paper. SCIS. https://wiredspace.wits.ac.za/server/api/core/ bitstreams/b97d9116-e2bf-4516-a0de-ca48b7e9dd7f/content

Walsham, G. (2012). Are we making a better world with ICTs? Reflections on a future agenda for the IS field. *Journal of Information Technology*, 27(2), 87–93.

Warschauer, M. (2004). *Technology and Social Inclusion: Rethinking the Digital Divide*. New York: MIT press.

Weitzberg, K. (2017). *We Do Not Have Borders: Greater Somalia and the Predicaments of Belonging in Kenya*. New York: Ohio University Press.

Weitzberg, K. (2020a). Biometrics, race making, and white exceptionalism: The controversy over universal fingerprinting in Kenya. *The Journal of African History*, 61(1), 23-43.

Weitzberg, K. (2020b). Kenya, thousands left in limbo without ID cards. Coda Story. https://www.codastory.com/authoritarian-tech/kenya-biometrics-double-registration/

Weitzberg, K., Cheesman, M., Martin, A., & Schoemaker, E. (2021) Between surveillance and recognition: Rethinking digital identity in aid. *Big Data & Society*, 8(1), 1–7.

WFP (2019a, February 5). Palantir and WFP partner to help transform global humanitarian delivery. https://www.wfp.org/news/palantir-and-wfp-partner-help-transform-global-humanitarian-delivery

WFP (2019b). A statement on the WFP-Palantir Partnership. https://medium.com/world-food-programme-insight/a-statement-on-the-wfp-palantir-partnership-2bfab806340c

Whitley, E. A., & Hosein, G. (2010). *Global Challenges for Identity Policies*. Basingstoke: Palgrave Macmillan.

Whitley, E. A., & Schoemaker, E. (2022). On the sociopolitical configurations of digital identity principles. *Data & Policy*, 4, 1–14.

Whitley, E. A., Gal, U., & Kjaergaard, A. (2014). Who do you think you are? A review of the complex interplay between information systems, identification and identity. *European Journal of Information Systems*, 23, 17–35.

Winner, L. (1980). Do artifacts have politics? *Dædalus*, 121–136.

World Bank (1998). *WDR 1998/99: Knowledge for Development*. Boston: World Bank and Oxford University Press.

World Bank and Nordic Trust Fund (2013). *Human Rights Impact Assessments: A Review of the Literature, Differences with Other Forms of Assessments and Relevance for Development*. https://documents1.worldbank.org/curated/en/834611524474505865/pdf/125557-WP-PUBLIC-HRIA-Web.pdf

World Bank Group (2015). *Identification for Development (ID4D) Integration Approach Study*. Washington, DC: World Bank Group.

World Bank Group (2019). *ID4D Practitioner's Guide*. https://documents.worldbank.org/en/publication/documents-reports/documentdetail/248371559325561562/id4d-practitioner-s-guide

World Bank Group (2021). *Identification for Development Annual Report*. https://documents.worldbank.org/en/publication/documentsreports/documentdetail/625371611951876490/identification-for-development-id4d-2020- annual-report

World Food Summit (1996). *Report of the World Food Summit*. https://www.fao.org/3/w3548e/w3548e00.htm

INDEX

A

Aadhaar, 2, 9, 29, 39, 61, 78, 86, 122, 132
  authentication, 40
  biometric identifier of, 20
  Central Identities Data Repository
    (CIDR), 54
  credentials, 19
  database, 22
  enrolment, 40, 70, 71, 73
  exclusions, 120–123
  Food Corporation of India's (FCI)
    database, 90
  functionality, 86
  grand design of, 44
  infrastructure, 40
  legal history of, 60
  physical presence, 43
  platform perspective, 32
  Public Distribution System (PDS), 40
  registration, 71, 73
  seeding, 52
  social protection schemes, 118
  verification, 31
Aadhaar-Based Biometric Authentication
  (ABBA), 51–53, 55, 56, 70, 71, 90, 135
  corruption, 56
  data aggregation, 91
  exclusion errors, 58
  functioning, 70
  identity fraud, 56
  Jharkhand, 55
  Karnataka, 57, 91, 135
  weighing-selling machine, 98
Aadhaar Enabled Biometric Attendance
  System (AEBAS), 29
Abnormal justice, 6
Above-poverty-line (APL), 16, 17, 94
Abuya, K., 127
Access Now, 132, 133, 158, 159
Accountability, 86, 91, 154
Adelmant, V., 113
Adverse digital incorporation, 34, 38
African Charter on Human and People's
  Rights (ACHPR), 149
Aggarwal, A., 77
Ahara, 60, 96
Akshaya telecentres, 142–144

Algorithmic fairness, 30, 154–156
Algorithmic opacity, 31
Allied Media Conference (AMC), 92, 146
Allu, R., 59
Alston, P., 68
Amnesty International (2020), 46
Andhra Pradesh, 142
Anganwadi, 121, 124
Anna Bhagya scheme, 94
Anti-poverty
  artefacts, 57–61
  programmes, 16, 44
  system, 20
Antyodaya Anna Yojana (AAY),
  16, 70, 94
Application Programming Interfaces
  (API), 20
Artificial intelligence (AI), 30, 31, 98, 139
Asylum seekers, 4, 10, 117
Asylum shopping, 100
Atick, J., 3
Authentication, 1, 5, 15, 20, 53, 54
  Aadhaar, 40
  biometrics, 22, 60
Authorisation, 1, 5, 18, 20
  conditionality of, 5
  fingerprint reader, 15
  ration disbursal, 6
Authoritarian surveillance, 31
Automated Fingerprint Identification
  System (AFIS), 63, 100, 101
Azraq camp, 80

B

Bailur, S., 23
Bashir, Y., 128
Baxi, P., 29
Bayly, C. A., 72, 73, 85, 87
Bee's case, 155
Below-poverty-line (BPL), 16, 17,
  70, 94, 119
  cash transfers, 26
  Chhattisgarh, 150
  essential goods, 27
  misclassification, 75
Better Than Cash Alliance (2021), 83
Bhoomi club, 143

Big data: A revolution that will transform how we live, work, and think (2013), 35
Big ID, 33, 132
Binarism, 8
Biometric artefact, 61–69
  biography of, 94–98, 94 (table), 95 (figure), 97 (figure)
  Kenya, humanitarian assistance to double registration, 61–65
  policing and partnerships, 98–106
  Uganda, digital ID and exclusion from entitlements, 66–69
Biometric Identification Management System (BIMS), 108
Biometrics, 1, 43
  Aadhaar, identifier of, 20
  assemblage, 139
  authentication, 22, 60
  humanitarianism, 102–106, 114
  machine, 1
  point-of-sale machine, 5
  recognition, 59
  records, 4
  technologies, 3
Blockchain-for-refugees, 78–81
Blockchains
  feature of, 78
  project, 44
  self-sovereign identity, 78
  as sociotechnical infrastructure, 139
Body scanners, in airports, 45
Border mafia, 52, 56
Border technologies, 98
Boundary resources, 20
Brazil, 74
Breckenridge, K., 32
Browne, S., 146
Bureaucracy, 29

C
Card, 15–18
Carswell, G., 60, 136–139, 145, 152
Cash entitlements, 9
Cash-for-work programmes, 44, 79
  beneficiaries of, 81
  women in, 79
Cash transfer system, 26, 38, 77
Central Identities Data Repository (CIDR), 54
Centre for Human Rights and Global Justice (2021), 55, 67
Centre of Global Development, 3
Chaudhuri, B., 44, 56

Cheesman, M., 44, 78–81, 85, 86, 139–141, 145, 154, 158
Chhattisgarh, 77, 149–151
Cioffi, K., 68, 113
Civil registration, 22
Clark, J., 21, 22
Collaboration on International ICT Policy for East and Southern Africa (CIPESA), 68
Colombia, 45
  Ingreso Solidario in, 81–85
  social protection system, 81
Colour-coding of ration cards, 19, 20
Compassionate repression, 109
Conditionality
  authorisation, 5, 20
  healthcare and cash entitlements, 9
  legal rights and entitlements, 7, 9, 110
Consolidated National Social Registry, 83
Constitutionality, 60
Coppi, G., 157
Co-production, 99
Corbridge, S., 15, 34, 73
Costanza-Chock, S., 8, 10, 45, 46, 92, 93, 104, 106, 107, 145, 146
COVID-19 from the Margins: Pandemic Invisibilities, Policies and Resistance in the Datafied Society, 81
COVID-19 pandemic, 45
  data visualisations, 117
  informational injustice, 44
  welfare delivery, 83
Cristiano, F., 157
Cukier, K., 35
Cusumano, M. A., 31–32

D
Dadaab, 63
Dahan, M., 3, 23
Dark matter vision, 46
Das, S., 7, 39, 41, 45, 93
Data activism, 115–118, 133–134, 160
  proactive, 117
  reactive, 117
Data aggregation, 91
Data availability, 35
Data-erasing artefact, 86
Datafication, 28, 29, 35, 37, 48
Datafiers, 37
Data for dignity, 85, 152–154
Data fumes, 35
Data infrastructures, 98–102
Data justice, 34–37, 160
  definition of, 6, 9, 36

design-related injustice, 45–47
digital identity and, 37–39, 41–47
distributive, 36
elements of, 38
framework, 39–47, 41 (figure)
informational injustice, 43–45
instrumental, 36
legal injustice, 42–43
procedural, 36
rights-based, 36
structural, 36
suitability of, 6
taxonomy of, 36
Data Justice (2022), 37, 47
Data Justice Conference, 37
Data Protection Act, 127
Data Protection Impact Assessment
    (DPIA), 127
Data revolution, 35
Debt-induced suicides, 75
Decentralisation, 33
Dencik, L., 6, 37, 47
De Neve, G., 60, 136–139, 145, 152
Denial of information, 74–77
Deregistration, 127–130
*Design Justice: Community-Led Practices to
    Build the Worlds We Need* (2020),
    45, 92
Design Justice Network, 92
Design-related injustice, 8, 10, 39,
    45–47, 89
  biometric artefact, 94–106
  dark side, 92–93
  definition of, 89–92, 90 (figure)
  digital ID, dark matter of, 106–110
Deterioration, 75
Developing nations, 38
Development-oriented identification, 3
Devereux, S., 4, 24, 58, 83
Dharna, 120
Dichotomy, 22
Digital authentication, 4
Digital data, 6
Digital evolution, 20
Digital humanitarianism, 102, 103
Digital ID, 1–5, 113
  advocacy for maternity entitlements,
      124–125
  algorithmic fairness and, 154–156
  architecture of, 5, 20–21, 21 (figure)
  authentication, 1, 5
  authorisation, 1, 5
  cash entitlements, 9
  component of, 3

dark matter of, 106–110
as a datafier, 28–30
data justice and, 37–39, 41–47
detrimental outcomes, 7
development and, 22–25, 23 (figure)
gateway to food, 118–125
healthcare, 9
as a mediator of surveillance, 30–31
as a platform, 31–33
resistance, 116–118
spread of, 2
Uganda, 66–69
Digitalisation of identity, 15–33
  card, 15–18
  development and, 22–25, 23 (figure)
  digital identity systems, architecture of,
      20–21, 21 (figure)
  foundational and functional identity,
      21–22
  principle, 18–20
  problematic link, 25–27
  three views of, 27–33
Digital performance, functions, 1, 5
Digital rights, 156–157
Digital social protection, 152–154
Digital wallets, 79
Digitisation, 4
Direct benefit transfers (DBT), 119
Direct cash transfers, 27
Discrimination, 6, 38
Distributive data justice, 36
Distributive justice, 6
Documentation practices, 18
Double registration, 43, 61–65
Drèze, J., 55, 56, 58, 75, 97, 137, 150
Dr Muthulakshmi Maternity Assistance
    Scheme (MAMTA), 125
Drones, 99
Dublin Regulation (1990), 99

E
Economic and Political Weekly, 75, 119
Economic vulnerability, 24
Ecosystem, 149
e-governance, 142
Empowerment, 30, 33
Engine Room, 114
Environmental Impact Assessments (EIAs),
    147
Epistemic control, 99
Error, 58
  exclusion, 4, 24
  inclusion, 4, 24
Essae Teraoka, 95

Ethnography, 80, 85, 137, 142, 145
EUMigraTool (EMT), 109
Eurodac, 46, 98–102
EuroMed Rights (2023), 98, 99, 102, 109
European Commission, 100
Excluded needy, 55, 91, 107
Exclusion errors, 4, 24, 26, 58, 70
Exemption register, 58
EyePay, 79, 86, 140, 141

F
Facebook, 35
Fair ID, 7, 10, 135–158
    algorithmic fairness and, 154–156
    conceptual apparatus, 152–158
    data for dignity, 152–154
    digital rights space, 156–158
    informational injustice, 137–139
    infrastructure justice, 139–151
    smart cards, in Tamil Nadu, 135–137
Fair-price shops, 17
FALCON analytical platforms, 105
Fassin, D., 108
Favouritism, 6, 38
FCI. *See* Food Corporation of India (FCI)
Financial and Accounting System
        (FIST), 96
Financial independence, 44
Fingerprinting, 5, 6, 15, 51, 136
    Aadhaar, 40
    Eurodac, 46
    readers, 20
    recognition, 71
Firmino, R., 108
Floridi, L., 154
Food activism movements, 118
Food Corporation of India (FCI), 16, 17, 90
Food Department, 16
Food distribution, problem of, 26
Food rations, 1, 5, 10, 16
Food security programmes, 16, 53
Foundational identity system, 21–22
Franz Edelman Award, 105
Fraser, N., 6, 7
Functional identity system, 21–22
Fundamental services, improved
        access to, 24
Fussy, P., 117

G
Garissa, 62
Gaza Strip, 157, 159
Gelb, A., 3, 21, 22, 23
Gender binary identification systems, 8

Global South, 47
Godowns, 96
Governmental schemes, 15
Government-to-person (G2P) payments,
        83
Great Integrated Household Survey, 153
Green Card Scheme, 57
Green Revolution, 16

H
Haki na Sheria, 62–65, 128, 129, 148
Harish Gowda, B. A., 94–96, 98
Healthcare, 9, 67
Heeks, R., 36, 43, 47, 78, 93
Household-based document, 19
Huduma Namba, 64, 126, 127
Humanitarian agencies, 2
Humanitarian assistance, 25, 31, 61–65
Humanitarian blockchain project, 86
Humanitarianism, 44, 46, 139
Humanitarian–private partnership, 46
Human Rights Impact Assessments
        (HRIAs), 145–149, 153
Human Rights Watch, 156
Hundal, H. S., 56, 59, 91, 95, 135,
        137, 138
Hunger deaths, 43

I
Iazzolino, G., 10, 31, 47, 108, 109
ID4Africa, 3
Identification, 1, 3, 5, 54
    biometrics, 59
    development-oriented, 3
    embodiment of, 2
    technologies, 4
Identification for Development (ID4D), 1,
        3, 28, 38
Identity-entitlement matching, 21
Inclusion errors, 4, 24, 26, 42, 58, 91
Inclusion of minorities, 24–25
Incorrect identification, 54
India, 2
    Aadhaar, 2, 9
    anti-poverty system, 20
    bureaucracy, 29
    documentation practices, 18
    Economic Survey (2015/2016), 25
    food grain production, 16
    International Institute for Information
        Technology Bangalore (IIITB), 2
    Jharkhand, 39, 54
    Karnataka, 39
    Ministry of Finance, 26

National Employment Rural Guarantee
 Act (NREGA), 142
National Sample Survey (NSS), 75
National Skills Registry (NSR), 29
Public Distribution System (PDS),
 16, 43, 53
Rajasthan, 53
ration cards, 18, 22
Right to Food campaign, 115,
 118–125
social protection system, 28
Tamil Nadu, 52, 55, 139, 149, 151
Telangana, 30
welfare history, 18
Indian Constitution, Article 21, 120
Indian Institute of Technology (IIT), 53
Indira Gandhi Matrutva Sahyog Yojana
 (IGMSY), 124
Informational injustice, 7–9, 39, 43–45,
 70–73, 137–139
blockchain-for-refugees, 78–81
COVID-19 pandemic, 44
enactment of, 44
information-erasing databases, 81–85
information orders and, 85–88
knowing, 72–73
opacity, 74–85
Information-erasing
artefacts, 44
databases, 81–85
machine, 78–81
technologies, 77–85
Information order, 72, 73
Information Systems (IS), 32, 46
Information Village Research Project
 (IVRP), 72
Infrastructure justice, 81, 139–151
anti-injustice ID artefacts, 149–151
dark matter, 144–151
deregistration, 127–130
Human Rights Impact Assessments
 (HRIAs), 145–149
materialities, 140, 141
subjectivities, 140
timescapes, 140, 141
user–provider interface, 141–144
Infrastructures of compassionate
 repression, 10
Ingreso Solidario, in Colombia, 81–85
Injustice, 5, 9. See also Specific types
design-related, 8, 10, 39, 45–47
informational, 7–9, 39, 43–45, 137–139
legal, 7, 39, 42–43
Instrumental data justice, 36

Integrated Case Management System
 (ICM), 105
Integrated Child Development Services
 (ICDS), 118, 123
International Covenant on Economic, Social
 and Cultural Rights (ICESCR), 149
International Human Rights Law Clinic, 148
International Institute for Information
 Technology Bangalore (IIITB), 2, 72
Interoperability, 4

J
Janani, A. P., 56
Janani Suraksha Yojana (JSY), 124
Jan Dhan Yojana, Aadhaar and Mobile
 numbers (JAM), 26, 76
Jharkhand, 39, 54, 55, 58
Jordan, 9, 79
automated cash transfers, 156
cash-for-work programme, 44
refugees, 44–45
Joshi, D., 154
Jumbert, M. G., 99
Justice
abnormal, 6
distributive, 6
as fairness, 6
infrastructure, 127–130
Justice Wadhwa Committee Report (2010),
 52, 58, 74, 96

K
Karnataka, 19, 25, 27, 39, 43, 59, 60, 77,
 122
Aadhaar-Based Biometric Authentication
 (ABBA), 57
Aadhaar enrolment in, 40
Anna Bhagya scheme, 94
compulsory Aadhaar linkage, 54
entitlement to food grains, 94 (table)
godowns, 96, 97 (figure)
Green Card Scheme, 57
point-of-sale machine, 51, 95 (figure)
ration dealers in, 40
Kenya, 10, 115, 148
double registration, 43, 61–65
humanitarian assistance, 61–65
human rights obligations, 149
Maisha Namba, 130
new ID, 125–127
refugee camps, 31, 62
Somali ethnicity in, 62
Kerala, 52, 118, 122
coloured ration cards, 20

legal injustice, 42
Ministry of Food and Civil Supplies, 18
Public Distribution System (PDS), 90
  (figure)
Ration Card Management System
  (RCMS), 19
ration cards in, 16, 19, 20
ration dealers, 17, 75
telecentres, 142
Kerala Rationing Collector, 18
Khera, R., 53, 59, 75, 77, 97, 98, 136, 150
Kidd, S., 67
Kitchin, R., 117
Knowing, 72–73
Krishna, S., 31

L
Langlois, G., 116
LC1 Chairperson, 66
Legal identity, 25
Legal injustice, 7, 39, 42–43, 113
  anatomy of, 52–54
  anti-poverty artefacts, 57–61
  biometric artefact, 61–69
  concept of, 42
  family matter, 51–52
  Kerala, 42
  numbers of, 54–57
López, J., 82–84, 152–155
López-Solano, J., 81
Loud No to Cash, 149
Low and middle-income countries, 2
Lyon, D., 30, 31

M
Madianou, M., 139
Madon, S., 32
Maisha Namba, 64, 65, 126, 130
Martin, A., 2, 46, 65, 103, 158
Martins, B. O., 99, 105, 106
Masiero, S., 7, 23, 39, 41, 45, 93
Mastodon, 35
Materialities, 140, 141
Maternity entitlements, 121, 124–125
Mayer-Schönberger, V., 35
Mediator of surveillance, 30–31
Milan, S., 10, 116–118, 132–134
Minorities, inclusion of, 24–25
Misalignment, 45
Mistrust, 4
Modular Open Source Identity Platform
  (MOSIP), 2
Mooij, J. E., 16, 95
Moses, R., 57

MOSIP. *See* Modular Open Source Identity
  Platform (MOSIP)
Mukhyanmantri Khadiyann Sahayata
  Yojana (MKSY), 151
Murakami Wood, D., 108
Muralidharan, K., 55, 56
Mutung'u, G., 64

N
Nair, V., 29
Napolitano, A., 98
National Administrative Department of
  Statistics (DANE), 153
National Aid Fund, 156
National Bureau of Registration, 63, 64
National Democratic Alliance (NDA), 40
National Employment Rural Guarantee Act
  (NREGA), 142
National Food Security Act (NFSA), 19, 20,
  53, 120, 121
National Identification and Registration
  Authority (NIRA), 66
National Identity Card (NIC), 66
National Identity Number (NIN), 66
National Identity Register (NIR), 66
National Integrated Identity Management
  System (NIIMS), 64
National Planning Department (NPD), 82
National Sample Survey (NSS), 75, 76
National Skills Registry (NSR), 29
National Social Security Fund (NSSF), 126
Ndaga Muntu, 66
New ID, 125–127
Nilekani, N., 2, 40, 54, 148
Non-digital artefact, 18
Non-priority households (NPHH), 19
Nordic Trust Fund, 146–148
Nutritious food, 121
Nyst, C., 5

O
Odisha, 125
Opacity, 44, 73
  denial of information, 74–77
  information-erasing technologies, 77–85
Orthodoxy, 2, 4, 25
  development, 3
  openly articulated, 2

P
Padyatra, 120
Palantir, 46, 65, 104, 106
Palestine, 157
Peace Research Institute Oslo (PRIO), 99

Pelizza, A., 46, 99, 100
Person's identity
  poverty status, 20
  ration dealer to check, 20
  uniqueness of, 3
Person's visibility, 23
Piecemeal pedagogies, 29
Platform, 31–33
  Aadhaar, 32
  characteristics, 31
  definition of, 31–32
Point-of-sale machine, 91, 95 (figure)
Political economy, 48
Political lives, 88
Poverty levels, food grain allocations, 74
Poverty status, 2, 17
  conflation of identity and, 22
  person's identity, 20
Power relations, 36
Pradhan Mantri Jan Dhan Yojana, 26
Prakash, A., 60, 94
Pre-Aadhaar system, 89
Prieto, A. M., 83
Prime Minister's Public Finance Scheme,
  India, 26
Priority households (PHH), 19
Proactive data activism, 117
Procedural data justice, 36
Processing Citizenship project, 100
Provider of identity, 23
Public Distribution System (PDS), 9, 16, 19,
  20, 22, 25, 26, 52, 55, 58, 71, 74, 75
  Aadhaar-based architecture, Karnataka,
    90 (figure)
  Aadhaar's incorporation, 40
  biometric data of, 39–40
  biometric identification, 44
  cash transfer rejection, 77
  cash transfer system, 26
  Chhattisgarh, 151
  direct cash transfers, 27
  essential goods, 16
  food basket composition, 53
  godowns, 97 (figure)
  goods, 90
  high offtake, in 1965–1990, 17
  informational injustice, 43
  Karnataka, 39, 59
  leakage of, 26
  supply chain, 52
  Tamil Nadu, 59, 135
Public scrutiny, 96
Public vigilance, 96
Puducherry, 72

Puri, R., 77, 149, 150
Puttaswamy, K. S., 60

Q
QR code, 135–137

R
Rajasthan, 53
Ramakumar, R., 96
Rao, U., 29
Ration Card Management System (RCMS),
  18, 19, 42, 144
Ration cards, 1, 16, 18, 19, 22, 54, 59, 74
  electronic emission of, 42
  ethnographic approach, 18
  Karnataka, 19
  Kerala, 16, 19, 20
Ration collection, 71
Ration dealers, 17, 18, 51, 75
  accountability, 91
  Karnataka, 40
  Kerala, 75
  physical verification, 20
Rawls, J., 6
Reactive data activism, 117
Refugees, 1, 2. See also Rohingya refugees,
    in Bangladesh
  blockchain, 78–81
  cash-for-work programmes, 79
  Jordan, 44–45
  Kenya, 31
  repatriation, 114
Registration of Persons Act, 67
Renken, J., 36, 43, 47
Renzi, A., 116
Resistance, 10, 113–134
Rice card, 152
Rice mafia, 52
Rights-based data justice, 36
Right to Food campaign, 118–125
Roberts, T., 157
Rohingya refugees, in Bangladesh,
  114, 158
Rutenberg, I., 64

S
Sabates-Wheeler, R., 4, 24, 58
Samagra Vedika, Telangana, 155, 156
Schoemaker, E., 29, 32, 33
Schrock, A. R., 116
Science and Technology Studies
  (STS), 99
Scientific objectivity, 26
Scott, J. C., 15

Secretary of Food, Civil Supplies and Consumer Affairs, 89
Sector transgressions, 104, 158
Securitisation, 2
Self-sovereign identity, 78
Senior Citizens' Grant (SCG), 67, 68
Seva kendra, 52, 54
Shakthi, S., 29
Shekhar, S., 47
Singh, S., 54
Sinha, D., 119, 122, 123
Sisbén IV, 82–84, 153, 154
Smart cards, in Tamil Nadu, 59, 60, 135–137
　elements, 136
　point-of-sale machine screen displays, 135
　QR code, 135–137
Social cash transfers, 2
Social Impact Assessments (SIAs), 147
Social justice, 48
Social protection system, 24, 27, 28, 39, 58, 81
Social vulnerability, 24
Societal engagement, 117
Sociotechnical infrastructure, 139, 140
Software Development Kits (SDK), 20
Solanki, A., 29
Solidarity, 113–134
Somalia, 62
Srinivasan, J., 72, 73, 85, 87
Sriraman, T., 18, 19, 29, 95
Srivastava, K. S., 154
Star, S. L., 140
Structural data justice:, 36
Subjectivities, 140
Subsidised rations, 27
Sugar card, 145
Superapps, 33
Surveillance capitalism, 114
Surveillance technology, 99
Sustainable Development Goals (SDGs), 3, 22, 35, 57
Swaminathan, M., 55, 91
Swaminathan, S., 75
Systematic marginalisation, 2
Systemic diversion, 75

T
Taluk Supply Office (TSO), 18
Tamil Nadu, 52, 55, 139, 149, 151
　Dr Muthulakshmi Maternity Assistance Scheme (MAMTA), 125
　Public Distribution System (PDS), 75

smart cards in, 59, 60, 135–137
sugar card, 145
Tapasya, Sambhav, K., 30, 155
Tarafdar, M., 93, 146
Taylor, L., 6, 7, 9, 35–37, 103, 105, 156, 158
Technology, use of, 23 (figure)
　improved access to fundamental services, 24
　improved humanitarian assistance, 25
　inclusion of minorities, 24–25
Telangana, 30, 154, 155
Telecentres, 142
Temporary ration cards, 96
Thatcher, J., 35
Timescapes, 140, 141
Transcontextuality, 132
Transformative social protection, 4, 67
Transnationality, 132
Transparency, 137, 138
Twitter/X, 35

U
Uganda, 9
　digital ID, 66–69
　exclusion from entitlements, 66–69
UIDAI, 71
Unfair ID, 5–8, 41 (figure)
Unique Identification Project (UID), 2, 29, 59
Unique Personal Identifier (UPI), 126
United Cash Transfer Program, 156
United Nations Economic Commission for Africa (UNECA), 72
United Nations High Commissioner for Refugees (UNHCR), 2, 62, 63, 129
　digital wallets, 79
　refugee registration, 78
United Progressive Alliance (UPA), 40
Universal Declaration of Human Rights (1948), 52
Unmanned Aerial Vehicles (UAV), 99
User–provider interface, 141–144

V
Van der Velden, L., 10, 116–118, 132–134
Van Veen, C., 113
Veeraraghavan, R., 141, 142
Velez, V. O., 84
Visualisation, 36, 37
Vulnerability, 22, 115
　Colombia, 82
　economic, 24
　Sisbén IV, 84
　social, 24

W
Weitzberg, K., 61–63
Western Balkans, 102
Whitley, E. A., 32
#WhyID campaign, 10, 115, 130–133
Winner, L., 52, 57, 69, 149, 153
World Bank, 3, 17, 28, 54, 82, 84, 113, 146

World Food Programme (WFP), 10, 46, 65, 102–106
World Food Summit (1996), 53

Z
Zero-balance bank account programme, 26

www.ingramcontent.com/pod-product-compliance
Lightning Source LLC
Jackson TN
JSHW080036010226
97517JS00010B/104